KINDERGARTEN

Adventures in Shapes and Space

**Units 6-10
Science, Technology, Engineering,
Art, Math and Designing in 3D**

STEAMStart
By Jeannie Ruiz

Kindergarten Adventures in Shapes and Space

Units 6-10
Science, Technology, Engineering,
Art, Math and Designing in 3D

Copyright 2020 by Jeannie Ruiz
Published by PRINTING FUTURES PRESS
Vancouver, WA

AUTHOR: JEANNIE RUIZ
EDITOR: WAYNE STRIPLING, JR.

All rights reserved. No part of this publication may be reproduced, stored in a retrieval system, or transmitted in any form for any reason, recording or otherwise, without the prior written permission of PRINTING FUTURES PRESS

ISBN 978-1-942357-24-7

Photo permissions and authors, illustrators, designers,
Photographers available for download at the URL.

http://www.lnsl.org
https://printingfutures.com/

iNSL DEVELOPMENT TEAM (aka Ten80 Foundation)
STEMWORKS EXEMPLARY PROGRAMS DEVELOPED BY TEN80 Foundation

Author Jeannie Ruiz led the team that developed STEAMStart as well as upper grade level programs that are included in the STEMWorks Database by Change the Equation, a non-partisan, CEO-led initiative to connect and align efforts to improve STEM learning. These programs met the highest standards for excellence and exemplary STEM project based programming through rigorous examination by WestEd, an independent nonprofit research organization.

RECOMMENDED BY REAL TEACHERS

We are thoroughly enjoying our STEAM lessons. The program is excellent! My students have loved making the Rhombis and getting to build with shapes and tape. It has been a big hit! Thanks for everything!
Alicia Stenard
Mater Christi in Albany School District

"We need more projects like this that carry on for a longer number of weeks. My teachers used scope and sequence suggestions, and the program ran after school for a semester. We could have worked on this a lot longer."
Karen James
Druid Hills Elementary in Charlotte, NC

"I'm excited for our students to get their hands on real STEAM projects. There isn't much out there for Kindergarten and First Grade."
Sandy Mettler
Fort Worth Schools in Dallas, TX

Adventures in Shapes and Space Units 6-10

EDITOR'S COMMENTS

To the Educators using STEAMStart,

First of all, thank you for taking the leap. Project-based learning requires a commitment on the part of the teacher. It can be messy. It can mean more planning and more time spent gathering odds and ends that make a project work. The time and energy you spend developing imaginative responses and ability to problem solve will pay off tenfold.

There is significant evidence in our own research that indicates a need for memorization and rote learning IN ADDITION TO LONG-RANGE PROJECTS and innovative problem-solving. A decent project uses the challenge as a springboard for planning skills-based lessons. A great project urges kids to seek skills and knowledge that you can offer because they want to succeed in solving the problem you've posed. Quick recall of facts is necessary for innovative thinking. Kids don't need to stop and look up simple facts or count on teddy bears when we want them to focus on collecting data and making reasonable decisions.

Projects need outcomes. They need number-driven goals that qualify success in the challenges. Let kids get creative and try things that will fail. Did I say to let them fail? Let them fail. Innovators fail all the time. Some of our greatest inventions are the result of absolute and profound failures. Learning to use the failure as a tool rather than an emotional trial is where students are headed. It takes a few falls before learning to ride a bike. They will make a few bouncy balls on the road to a great slimy wall-walker.

The question arises as to why we chose to call students "designers" rather than "engineers" or "scientists." Designers incorporate a wide variety of communication skills as well as industry expertise. Because communication of ideas among students is integral to the success of any project based curriculum, the students become Rhombi's designers.

Our team celebrates the amazing lessons we learn from you. When you develop a new technique or unique approach to the material, let us hear from you. The resource site grows stronger as teachers share their ideas with each other.

Sincerely,

Jeannie Ruiz
Director of The International STEM Leeaague

DESIGNERS, WILL YOU HELP ME?

Designers collect, analyze, and synthesize data guided by specifications.

Designers make clear and concise recommendations using drawings, verbal descriptions, and math models.

The Industrial Designers Society of America's definition of industrial design describes design as "the professional service of creating and developing concepts and specifications that optimize the function, value and appearance of products and systems for the mutual benefit of both user and manufacturer."

Rhombi asks her friends to offer assistance in design. Students collect, analyze and make data-based decisions. They must share their concepts verbally, using drawings, and through models (scale and math).

RHOMBI'S ADVENTURES IN 3D CONTENTS

ACTIVITY		READ ALOUD
RHOMBI'S PLAYHOUSE *PAGE 19*		
WINDSOCKS	24	WIND ON THE WINDOWS
DESIGN	28	ACCIDENTAL INVENTIONS
STRUCTURES	32	THE TINY HOUSE
CUBISM	36	CUBE WORLD
PERSPECTIVE	40	DO YOU KNOW QUADRILATERALS
ROOF ROOF *PAGE 59*		
WATER /EROSION	64	WATER IS AS WATER DOES
PYRAMIDS	68	CASE OF THE DISAPPEARING SAND
ROOF HEIGHT	72	CIRCLE, TRIANGLE, SQUARE
MATERIALS	76	I LOVE HOMES
EROSION	80	THE CAMEL AND THE PYRAMID
PET FINDS A HOME *PAGE 97*		
HABITATS	102	MEGHAN FINDS A PURRL
ATTRIBUTES	106	RHOMBI LIVED IN A ZOO
SHAPES	110	IT ALL ADDS UP
PATTERNS	114	M.C. ESCHER'S WORLD
SCALE	118	HOW BIG IS A GUINEA PIG
HAPPY BIRTHDAY, PET *PAGE 136*		
TOOLS	140	THAT'S BANANAS
DIGITAL SCALES	144	WAXING COLORFUL
PERSPECTIVE	148	STRENGTH IN NUMBERS
MOSAICS	152	SQUARE IN THE MIDDLE
LENGTH	156	COMIN' UP A CLOUD
RHOMBI'S CANDIES *PAGE 170*		
EXTENSION		

INTRO TO STEAM

6	UNIT OVERVIEWS
8	LONG RANGE PROJECTS
9	GEOMETRY IS THE THREAD
10	PROJECT BASED LEARNING
11	GREAT STEAM CHECKLIST
12	CONFIDENT STEAM STUDENTS
13	ADULT MISCONCEPTIONS
14	DEFINING STEAM
15	DEVELOPING A STEAM MINDSET
16	BUILDING a CLASSROOM
17	USING THE MATERIALS

APPENDIX

128	NETS
130	POLYGON TRACING PAGES
165	MINDBUGS MEASUREMENT
175	NGSS CORRELATIONS
176	PRE AND POST ASSESS
186	VOCABULARY
188	COMMON CORE CHARTS
191	PLANNING PAGES
193	WORKS CITED

RHOMBI'S ADVENTURES IN 3D
UNIT OVERVIEWS

RHOMBI'S PLAYHOUSE (Play with "Sock It to Me")   19

Designers will help Rhombi design and build a playhouse.

SCIENCE: Build an anemometer to measure wind speed and chart wind direction. | 24 |

TECHNOLOGY: Design a wind-proof wall with classroom materials. | 28 |

ENGINEERING: Develop mathematical nets for a cube. | 32 |

ART: Create a geometric image using shapes. | 36 |

MATHEMATICS: Make perspective drawings to count cubes in a stack. | 40 |

ROOF ROOF (Play with "Learn to Juggle.")   59

Designers will help Rhombi add a water resistant roof to her playhouse.

SCIENCE: Test porous surfaces, and graph the size of puddles. | 64 |

TECHNOLOGY: Visit a real pyramid in the virtual world with a web-based Giza tour. | 68 |

ENGINEERING: Build and collect data on tall pyramid structures. | 72 |

ART: Make clay bricks to bake in the sun. | 76 |

MATHEMATICS: Use the clay pyramid to record erosion, and make predictions about size. | 80 |

RHOMBI'S ADVENTURES IN 3D
UNIT OVERVIEWS

PET FINDS A HOME (Play with "Hoop and Hollar") — 97

Designers will help Rhombi create a home for Pet.

SCIENCE: Create a classroom zoo habitat with a focus on shapes used in structures. — 102

TECHNOLOGY: Students design their "wild selves" online. — 106

ENGINEERING: Make a floor that holds a specified amount of weight. — 110

ART: Create your own tessellations. — 114

MATHEMATICS: Start a giant class bulletin board for categorizing size and scale — 118

HAPPY BIRTHDAY, PET (Play with "Hopscotch") — 136

Designers will help Rhombi create a space to celebrate Pet's birthday.

SCIENCE: Prepare a super banana treat for Pet. — 140

TECHNOLOGY: Use the digital scale to begin developing a sense of mass and weight. — 144

ENGINEERING: Use papers to devise a column that supports 1 pound of material. — 148

ART: Use tiles to create a mosaic. — 152

MATHEMATICS: Use string and measuring tapes to create mathematical art projects. — 156

RHOMBI'S CANDIES — 170

Designers will take part in an authentic assessment as they create a market candy cart.

RHOMBI'S ADVENTURES IN 3D
MATH MODELING IN K-2 SETTINGS

DEFINING THE MATH MODEL

Mathematics can be used to "model", or represent, how the real world works. A model is not the real thing, but it should be accurate enough to be useful. Math models connect abstract numbers to real experience. Math models save time, supplies, and aggravation. For example, Rhombi needs to build a full-scale cart. Since the class cannot build the full sized vehicle, it will develop a small scale version and extrapolate from there. They use their background knowledge to begin the math model.

MOVE FROM EXPERIENCE TO SYMBOLS

The study of science and math started in an effort to understand actions that could be observed directly. In order to make science and math a useful performing art, instruction must start with real experience and gradually build toward an abstract representation. Doing things in the reverse chronological order produces nearly useless results.

KNOW WHERE THE LESSON IS HEADED

Education separates experience from the learning process. We could learn a few things from our own history. Math development progressed along its slow, continuous track: caveman to calculus. Here, in the late 20th century, we expect second graders to operate at abstract levels using a symbolic language that took mathematicians thousands of years to develop. Educators are asked to start with the algorithm –the abstraction- and work backwards to the concrete. We wonder why math is such a dreaded subject. Neither the teacher nor the student knows where the symbols are supposed to take them.

STEAM USES NATURAL EXTENSIONS

Abstract models of STEAM phenomena should grow as a natural extension of what is felt by direct experience. The abstraction must be an immediate and obvious representation of what students already know at every level. Drive a radio controlled car. Build the shapes as you memorize them. Investigate, collect data and analyze results. From reality to abstraction, STEAM learning needs more teachers who understand that doing more means going further, faster.

DEFINE YOUR MATHEMATICAL MODEL

Identify the problem, define the terms in your problem, & draw diagrams where appropriate. Begin with a simple model, stating assumptions that you make as you focus on particular aspects of the phenomenon. Identify important variables and constants. Determine how they relate to each other. Develop the equation(s) that express relationships between variables & constants.

Rhombi's Adventures in 3D Project Based Learning

Learning shapes is great. Doing short-term interactive projects that involve shapes is even better. Turning those projects into inquiry-based challenges takes things a step further. Involving students in long-range projects that require multiple steps, depth of knowledge, and true integration of science and math is the goal. The National Science Foundation funded Innovative Technology Experiences for Teachers (ITEST) research that developed a list of criteria for excellence in project based learning.

CRITERIA FOR EXCELLENCE IN PROJECT BASED LEARNING

- Focus on activities that motivate students to develop interest and skills in STEM careers.

- Provide intensive engineering experience of a long duration (academic year and/or summer proportional to the nature of the program).

- Use technology tools and hands-on learning.

- Involve industry mentors, scientists, and engineers.

- Incorporate students' self-regulated learning within specific engineering design experiences.

- Offer innovative, high interest STEM activities.

- Involve participants in the designing and building of artifacts.

- Use scientific integration and/or the engineering design process.

- Engage students in project-based and/or problem-based learning.

- Develop participants' technical skills (as appropriate to the project).

- Make explicit connections between (STEM) academics and project.

- Build in success experiences and opportunities to learn from failure.

- Explore technical careers from technicians to engineers and engineering management.

RHOMBI'S ADVENTURES IN 3D
GREAT STEAM TEACHER CHECKLIST

1. DO MORE THAN IMPART KNOWLEDGE Children will teach themselves if offered the opportunities to explore and problem solve. Teach by posing questions. Teach by developing curiosity. Teach by guiding learners with interactive experiences. Teach to learn. Learn as much as you teach. Adults often take so much control of the teaching process that children resort to mimicry as a way of pleasing their mentors. Every child learns at a unique pace and with a unique set of circumstance; therefore each child requires a unique path. If educators view learning as a problem-solving process, learning becomes an individual journey.

2. DISLODGE THE MISCONCEPTIONS Every child enters school with a unique set of skills and preconceptions about learning. Every child has already formed opinions based on observation, mimicry, and the outcome of challenges (how were successes and failures handled by adults). Some facts that lodge in a child's mind are erroneous. Other ideas are keen observations of the natural and technological world. Facilitate learning. Increase knowledge. Guide inquiry. Open minds to the possibilities. Offer multiple opportunities to learn and approaches to problem solving as a way of addressing misconceptions.

3. USE WHOLE GROUP SPARINGLY There are rare teachers with the ability to engage learners as a group. Generally, this interaction has a fast-paced Q&A quality and involves a points system or prizes. The up-side is that learners are involved as a group and learn courteous behavior. The down-side is that educators are rarely able to get individual feedback or offer reinforcement / stimulation. Even technologically based feedback systems are an impersonal way to discover a student's strong and weak areas.

4. LEARN MORE WITH SMALL GROUPS Teachers that interact in small groups with students on a regular basis know far in advance of testing exactly what the outcomes will be. Anxiety lessens for teachers and learners. Differentiation becomes possible, and students no longer "tune out" the instructor when the instructor becomes a facilitator. It's hard to ignore the teacher when she's sitting at your table asking questions and challenging assumptions. It's difficult to avoid detection if the investigation requires active participation. For the learner, there is no more "tuning out" or sleep-walking through the day.

5. ASSESS STUDENTS REGULARLY There are no surprises when learners take part in authentic and active assessment with reinforcement for some and mental stimulation for others. Great teachers know well ahead of time how students will perform on standardized tests.

RHOMBI'S ADVENTURES IN 3D
SUCCESSFUL STEAM STUDENTS

Children View Themselves as Scientists in the Process of Learning.
1. They look forward to doing science.
2. They demonstrate a desire to learn more.
3. They seek to collaborate and work cooperatively with their peers.
4. They are confident in doing science;
   they demonstrate a willingness to modify ideas, take risks, and display healthy skepticism.

Children Accept an "Invitation to Learn" & Readily Engage in The Exploration Process.
1. Children exhibit curiosity and ponder observations.
2. They move around selecting and using the materials they need.
3. They take the opportunity and the time to "try out" their own ideas.

Children Plan and Carry Out Investigations.
1. Children design a way to try out their ideas, not expecting to be told what to do.
2. They plan ways to verify, extend or discard ideas.
3. They carry out investigations by: handling materials, observing, measuring, and recording data.

Children Communicate Using a Variety of Methods.
1. Children express ideas in a variety of ways: journals, reporting out, drawing, graphing, charting, etc.
2. They listen, speak and write about science with parents, teachers and peers.
3. They use the language of the processes of science.
4. They communicate their level of understanding of concepts that they have developed to date.
   Children Propose Explanations and Solutions and Build a Store of Concepts.

Children offer explanations from a "store" of previous knowledge.
(Alt Frameworks, Gut Dynamics).
1. They use investigations to satisfy their own questions.
2. They sort out information and decide what is important.
3. They are willing to revise explanations as they gain new knowledge.

Children Raise Questions
1. Children ask questions (verbally or through actions).
2. They use questions to lead them to investigations that generate further questions or ideas.
3. Children value and enjoy asking questions as an important part of science.

Children Use Observation.
1. Children observe, as opposed to just looking.
2. They see details, they detect sequences and events; they notice change, similarities and differences, etc.
3. They make connections to previously held ideas.

Children Critique Their Science Practices.
1. They use indicators to assess their own work
2. They report their strengths and weaknesses.
3. They reflect with their peers.

"Inquiry Based Science: What Does It Look Like?" Connect Magazine (published by Synergy Learning), March-April 1995, p. 13.

RHOMBI'S ADVENTURES IN 3D
ADULT MISCONCEPTIONS

These misconceptions were identified based on in-depth interviews with early childhood teachers about the key issues in early mathematics education as well as researchers' experiences in teaching early childhood students, conducting workshops with early childhood teachers (Ginsburg, Jang, Preston, VanEsselstyn & Appel, 2004; Ginsburg et al., 2006), working with them in early childhood classrooms, and engaging in informal conversations with them. The study also based descriptions of the myths on available research literature (Ginsburg, Lee & Boyd, 2008). The nine misconceptions defined by Joon Sun Lee of Hunter College, The City University of New York and Herbert P. Ginsburg of Teachers College, Columbia University are:

1. Young children are not ready for mathematics education.
2. Mathematics is for some bright kids with mathematics genes.
3. Simple numbers and shapes are enough.
4. Language and literacy are more important than mathematics.
5. Teachers should provide an enriched physical environment, step back, and let the children play.
6. Mathematics should not be taught as stand-alone subject matter.
7. Assessment in mathematics is irrelevant when it comes to young children.
8. Children learn mathematics only by interacting with concrete objects.
9. Computers are inappropriate for the teaching and learning of mathematics.

These misconceptions often interfere with understanding and interpreting the new recommendations on sound early childhood mathematics education, and become subtle (and sometimes overt) obstacles to implementing the new practices in the classrooms (Richardson, 1996).

A WORD FROM TEN80 TO TEACHERS
Become aware of your personal biases. Just because you had poor math instruction does not mean math is hard.

The language of math cannot be taught with the same symbolic approach as the language of reading and writing. Because everyone talks and practices the art of communication, teachers assume that words are the only way to demonstrate proficiency in every field including math and science. Math and science are performing arts that must be demonstrated in performance. Math and science must use pictures and numbers depicting an actual experiment that is a test of the student's abstract model of some aspect of reality. In this way, students build their own understanding of the patterns that underlie math and science. (Ten80 2002)

RHOMBI'S ADVENTURES IN 3D
DEVELOPING A STEAM MINDSET

While STEAM (Science, Technology, Engineering, Arts, Mathematics combined) is a new buzz word in the field of Education, it was already a well established practice before Mozart gave his first concert as a child prodigy (Salzburg, Austria, 1761).

Learning to incorporate math modeling, grounding the symbolic nature of numbers in reality, and turning the language of mathematics into an integral part of ALL subjects are the true keys to successful STEAM classrooms. Developing confidence in the understanding of size, scale, number, place value, data collection, data analysis and real problem-solving reasonableness is necessary for the creation of a true STEAM culture.

Start each day with a measurement and estimation challenge. As students interact with blocks, toys, manipulatives and other classroom objects, connections should be drawn to their counterparts in the real world.

Ex: Oh, Thomas, that is a 3ft tall tower you've built. Did you know that the Tower of London is 89ft tall? You could tell us a really great story about that tower. Renoir painted towers. Let's look at some of his online. What shapes do you see in the images? Maybe you could paint a Renoir style tower to illustrate your story about the scale model 3ft tower you've designed. If we had more blocks, what else would you add to the tower? A drawbridge? Wow, let's look at the design of a drawbridge's pulley system.

Expecting that level of interaction is realistic when learning centers are guided by the teacher. If all students will be using blocks, you can bet they'll build towers. Anticipating student interests allows the teacher to prepare for verbal interactions that lead to STEAM experiences.

Project-based learning dictates that teachers take the long view in planning. Begin with what you need to accomplish in order to meet some challenge (like addressing Common Core and Next Generation standards). What project will allow natural extensions to your subject matter but also draw kids in as far as interest level? Making the project tie in to real life is also advisable. Do you like to quilt? Incorporate that into your classroom as a tie to measurement, geometry, patterns, history, writing, etc. The project might be designing and making quilts for local shelters. Get creative. In the case of STEAMStart, we've given you the jump start. Now, you get to extend the project to other aspects of learning in your curriculum.

RHOMBI'S ADVENTURES IN 3D
DEFINING STEAM

STEAM (science, technology, engineering, arts, mathematics) is just now becoming an accepted acronym among teachers and is seen as a somewhat mysterious –almost magical- fix for all the problems American schools and businesses are facing. Perhaps we should start with what STEAM is not:
- STEAM is not a conglomeration of lessons culled from five subject areas.
- STEAM is not something you can cover in a unit or two every year.
- STEAM is not a 5-day or 10-day curriculum piece.
- STEAM cannot be delivered in the traditional pencil-and-paper method by a sage on the stage standing behind the classroom lectern or seated behind a grand old wooden desk. There are no "Buehler..? Buehler..?" moments in a STEAM environment.

(reference the movie Ferris Buehler's Day Out)

SO WHAT IS STEAM?
STEAM is a philosophy of teaching, learning and working. STEAM is more than just the sum of the words in its acronym.

In other words, do real stuff to find real answers to real problems.

STEAM IS PRETTY SIMPLE TO STEAM PROFESSIONALS
What has become a raging debate among educators seems pretty simple to the professionals that actually work in STEAM fields. Academicians will spout research from their PhD thesis and published articles in journals, etc. Ask a scientist, a mathematician, an engineer or a programmer how to prepare kids for a STEAM workplace and they'll tell you to make kids DO something.

SET UP CLASSROOMS FOR DISCOVERY
While activity centers are typical in kindergarten settings, they have all but disappeared by second grade. Bring them back. Small group interaction is optimal for STEAMStart as it is for any STEAM program. Why do you want to use stations or centers in the classroom?
- Simplifies planning - plan over a week rather than day by day...
- Offers opportunities for more cohesive lessons...
- Allows teachers to hold small group sessions daily...
- Encourages teamwork...
- Simplifies classroom management... *
- STEAM increases the ability to draw connections among various concepts.

* Reference   Harry Wong and The Effective Teacher
HTTP://www.effectiveteaching.com

STEAMStart copyright 2020 by Jeannie Ruiz All Rights Reserved

RHOMBI'S ADVENTURES IN 3D
BUILDING STEAM CLASSROOMS

According to corestandards.org, "The Common Core State Standards provide a consistent, clear understanding of what students are expected to learn, so teachers and parents know what they need to do to help them. The standards are designed to be robust and relevant to the real world, reflecting the knowledge and skills that our young people need for success in college and careers. With American students fully prepared for the future, our communities will be best positioned to compete successfully in the global economy."

Support Common Core Standards and Learning Goals

STEAM PROJECT STATION
Support the ongoing STEAM project with a permanent station and construction materials. Recycled and upcycled materials should be available for use along with tools, paper, pencil and data collection opportunities.

ART, MUSIC, DRAMA
Encourage discussion of STEAM subjects through the arts as well as investigations involving art inventions, art history, genres, musical instruments and sound, etc. Create a "stage" where students give skits, give presentations, teach lessons and make up songs about classroom learning.

BLOCK PLAY
Keep blocks available all the time along with small scale toys at different scales. Add books about world architecture, habitats, and building projects.

NONFICTION LIBRARY
Facilitate the transition from fiction and storybooks to gleaning facts from informational text. Fill your library with fiction AND non-fiction. Encourage students to write informational text and to keep journals for science and math.

STEAM CHARACTER ROLE PLAY
Most kindergarten classrooms include an area for dress-up and pretend. How many first and second graders would still benefit from this type of role-play? Shoot for higher level characterization with lab coats, construction hats, medical scrubs, artist's smocks, etc.

RHOMBI'S ADVENTURES IN 3D
GEOMETRY IS THE THREAD

The word "geometry" comes from the Greek words for "Earth" and "measure." Geometry was first used to measure and chart the length, area and shape of land surfaces.

GEOMETRY is the single math thread that runs through Common Core from Kindergarten to 12th grade. STEAMStart promotes STEAM and the Art of STEM through an authentic experience in engineering problem solving built on a platform of geometric constructions, analytical geometry, and engineering design. Math is a gateway to success in professional fields: speaking the language of mathematics fluently means that students may define their career paths by choice, not resignation.

~ Jeannie Ruiz

STEAMStart Developer and Director of STEM Ten80 Foundation

THE MATH GENE
Contrary to popular belief, no one is born with a genetic predisposition toward math. There is no "math gene" in humans (Keith Devlin, The Math Gene); mathematics is a learned skill for everyone just like basketball, playing the French Horn, or ballet. Humans have brains which are already wired to enable math acquisition (Stanislas Dehaene, Number Sense 1997). Infants, chimps, dogs, and birds can recognize groups of objects up to three and decide that an object has been taken away or added. That is not "doing math," but it is a qualitative foundation on which to build understanding. Math instruction in school develops the initial qualitative sense of space, time, and quantity into the descriptive -and highly useful- language of mathematics. Children should learn all things in context.

MULTI YEAR FLUIDITY
Geometry as a foundation allows the educator to implement multi-year projects that students enhance year after year. The essence of the project is still fluid, while monthly, weekly, and daily objectives are met for the grade level.

NEXTGEN SCIENCE STANDARDS - STRUCTURE AND FUNCTION
Engineering and Design: Students should understand that "the shape and stability of structures of natural and designed objects are related to their function(s).
(K-2-ETS1-2)."

RHOMBI'S ADVENTURES IN 3D
USING THE MATERIALS

ENGAGE Each activity uses a read-aloud book to jump start the investigation and engage students.

DISCUSS Establish mindbugs with an introductory discussion for each investigation. The MindBug is a misconception in student thinking; procedural understanding that worked in one situation but is incorrectly applied to other skills. Like a computer virus, MindBugs spread through a student's system.

Example In Science: Children often think that air weighs nothing because they see scales measuring zero from an early age.

Example in Math: Students line the numbers up on the right to add. This works until 5th grade when they attempt to add decimals.

EXPLAIN Each activity includes a brief background for the purpose of introducing the subject matter.

MOVE Get students up and moving to activate that extremely useful muscle memory.

INVESTIGATE

Although each EXPLORE activity and investigation stands alone, together they also build knowledge required to meet the long-range project. In Module 1, students will -ultimately- be challenged to help Rhombi prepare for a trip.

ELA Build skills in communication, writing and speaking.

Rhombi's Adventures in 3D

RHOMBI'S PLAYHOUSE

UNIT 6

RHOMBI'S ADVENTURES IN 3D
RHOMBI'S PLAYHOUSE

RHOMBI'S PLAYHOUSE

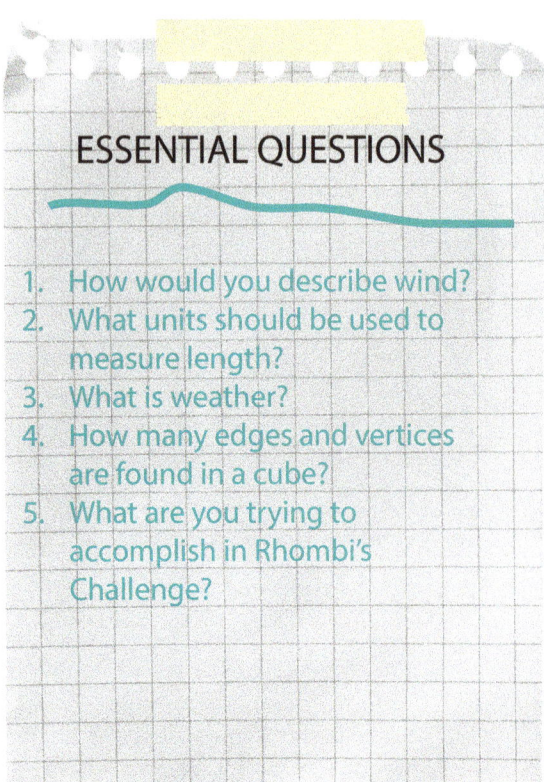

ESSENTIAL QUESTIONS

1. How would you describe wind?
2. What units should be used to measure length?
3. What is weather?
4. How many edges and vertices are found in a cube?
5. What are you trying to accomplish in Rhombi's Challenge?

Focus on squares and cubes. Construct an anemometer and create a windy record. Investigate mistakes that worked. Develop a solid wall seam using adhesives and materials. Take a closer look at geometric shapes in real life, and gain a better perspective in math.

Students will use basic measurement, understanding of number, place value, data collection, and data application in problem solving. Two-dimensional shapes will become three-dimensional shapes with a purpose.

Rhombi's playhouse is pretty rickety. She needs assistance in designing and building a new, strong, wind-resistant home. Students will collect data, design and construct a scale model that withstands the hair-dryer test.

Weather is about local changes in temperature, rainfall, wind speed and other factors.

A cube can be made from 6 squares.

Two-dimensional polygons / shapes can be combined to create many three-dimensional shapes.

weather
day to day conditions of a particular place

invention
new device, method, or process

perspective
way of representing 3D objects in 2D

skyscraper
a tall building with many stories

RHOMBI'S PLAYHOUSE

Rhombi needs a new cube playhouse to withstand windy weather.

COMMON CORE CONNECTIONS

Cubism was created by Pablo Picasso and Georges Braque in Paris between 1907 and 1914. Vauxcelles called the geometric forms in Braque's abstracted works "cubes." Cubist works emphasize the flat two-dimensional nature of canvas. Cubists reduced and fractured objects into geometric forms. Sculptors included Alexander Archipenko and Raymond Duchamp-Villon.
source: http://www.metmuseum.org

Interactive Whiteboard with **Polygon Sort**
Http://alturl.com/n9sg7

Play to Learn with **Shapes Games**
http://pbskids.org/games/shapes/

NEXT GEN SCI CONNECTIONS

A **CUBE** can be folded using 11 different possible nets (patterns for folding). In geometry, a cube is a three-dimensional solid object bounded by six square faces, facets or sides, with three meeting at each vertex. The cube is the only regular hexahedron and is one of the five Platonic solids. The cube is also a square parallelepiped, an equilateral cuboid and a right rhombohedron. It is a regular square prism in three orientations, and a trigonal trapezohedron in four orientations.

Faces	6 Sides
Vertices	8 Corners
Edges	12

Many students use cubes to learn multiples of ten. A single cube represents "1." 10 singles makes a row of 10. A row of ten cubes laid side by side becomes "100." "1000" is represented by stacking rows of ten to form a cube.

UNIT STEAM ACTIVITIES

The English word "cube" comes from Arabic "k'ab", meaning cube. Chinese, Indian and Islamic scholars were hard at work making contributions to mathematics during Europe's Dark Ages (when intellectual endeavor stagnated). In the 1100's and through the early 1200's, trade expanded the practical need for math. Fibonacci -the first significant mathematician in Europe in more than 1000 years- introduced Hindu-Arabic numerals and geometric vocabulary to Europe.

How do you square a cube? How do you cube a cube? How do you multiply a cube by 10? While using the block system to teach counting, multiples and base ten seems reasonable, beware that children can be confused by the 3-dimensional representation.

Swat the MindBug: When teaching with the cube system, also count using many other options. Bag rice or beans. Use a number line.

Read About STEAM

SCIENCE

Feel the Wind
Arthur Dorros

W is for the Wind
A Weather Alphabet
Dr. Suess

TECHNOLOGY

The Industrial Revolution Build It Yourself
Carla Mooney

Mistakes that Worked
Charlotte Jones Degen

ENGINEERING

How a House Is Built
Gail Gibbons

Olivia Builds a House
Maggie Testa

ART & MUSIC

Museum Shapes
Metropolitan Museum of Art

I Spy Shapes in Art
Lucy Micklethwaite

MATH

Cubes, Cones, Cylinders and Spheres
Tana Hoban

Shapes Around You: 3-D Shapes
Julia Wall

RHOMBI'S PLAYHOUSE
CURRICULUM CONNECTIONS

COMMON CORE CONNECTIONS

ELA/Literacy
RI.K.1 With prompting and support, ask and answer questions about key details in a text. W.K.1 Use a combination of drawing, dictating, and writing to compose opinion pieces in which they tell a reader the topic or the name of the book they are writing about and state an opinion or preference about the topic or book. W.K.2 Use a combination of drawing, dictating, and writing to compose informative/explanatory texts in which they name what they are writing about and supply some information about the topic.

Mathematics
MP.2 Reason abstractly and quantitatively. MP.4 Model with mathematics. K.CC.A Know number names and the count sequence. K.MD.A.1 Describe measurable attributes of objects, such as length or weight. Describe several measurable attributes of a single object. 1MD.A.1 Describe measurable attributes of objects, such as length or weight. Describe several measurable attributes of a single object.

NEXT GEN SCI CONNECTIONS

ETS1.A: Defining Engineering Problems
A situation that people want to change or create can be approached as a problem to be solved through engineering. Such problems may have many acceptable solutions.

1-ESS1-1, 1-ESS1-2 Patterns in the natural world can be observed, used to describe phenomena, and used as evidence.

ESS2.D: Weather and Climate
Weather is the combination of sunlight, wind, snow or rain, and temperature in a particular region at a particular time. People measure these conditions to describe and record the weather and to notice patterns over time. (K-ESS2-1)

K-ESS2-1. Use and share observations of local weather conditions to describe patterns over time.

UNIT S.T.E.A.M. ACTIVITIES

Rhombi Audio Download / Video (available December 2014)

UNIT 1:	Rhombi's Playhouse: Science:	Wind
UNIT 1:	Rhombi's Playhouse: Technology:	Design
UNIT 1:	Rhombi's Playhouse: Engineering:	Patterns
UNIT 1:	Rhombi's Playhouse: Art & Music:	Cubism
UNIT 1:	Rhombi's Playhouse: Mathematics:	Perspective
UNIT 1:	Rhombi's Playhouse: ELA:	Read and illustrate "Rhombi's Playhouse."

RHOMBI'S NEW HOME
STEMVESTIGATION: WIND

READ	**Wind in the Windows**
DISCUSS	What do we already know about wind? Describe wind's motion during various types of weather.
EXPLAIN	Wind is air in motion powered by the sun. Wind is described based on where the wind started. "Calm air" moves less than 1 mile per hour. "Hurricanes" cause violent destruction at 74 or more miles per hour.
MOVE	Simulate wind. Students wearing warm colors act as the warm air. Students in cool colors act as the cooler air. As the warm group moves out of an area, the cool group moves in waving fans.

INVESTIGATE	
How hard is the wind blowing? Build a weather vane to chart the wind direction. Start a data collection project that could last all year. Include additional elements		
ELA	Read and Explore "W is for the Wind" A Weather Alphabet. Create alphabet books!.

SCIENCE

RHOMBI'S PLAYHOUSE
STEMVESTIGATION: WINDSOCK

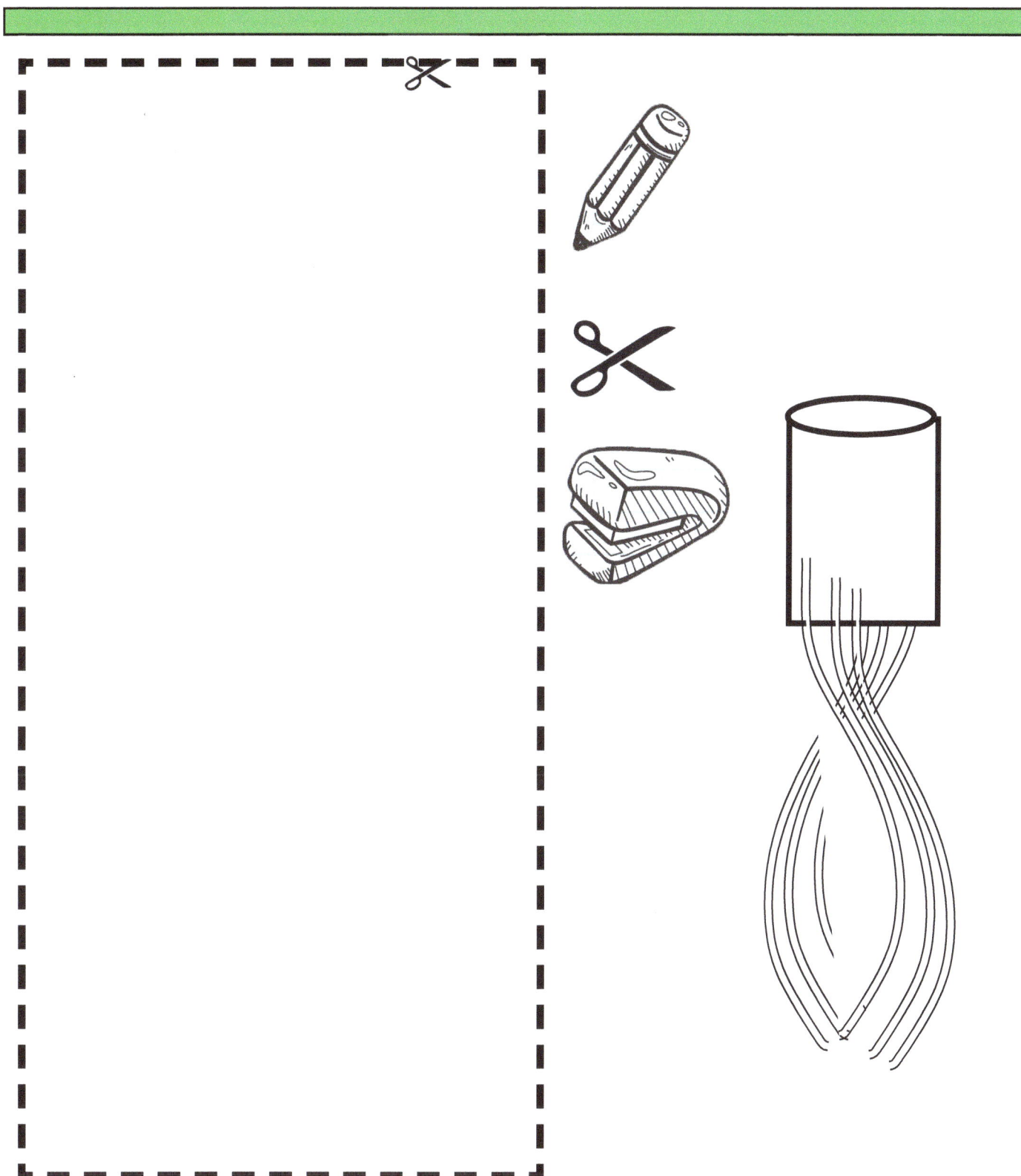

Wind on the Windows
Explore weather in science.

WORDS FOR WIND

CHINOOK
westerly wind off the eastern side of the Rocky Mountains

SANTA ANA
easterly towards Southern California

SCIROCCO
southerly from North Africa to southern Europe

MISTRAL
northwesterly from central France to Mediterranean

MARIN
southeasterly from Mediterranean to France

BORA
northeasterly from eastern Europe to Italy

GREGALE
northeasterly from Greece

ETESIAN
northwesterly from Greece

LIBECCIO
southwesterly towards Italy

26

Wind on the Windows
Explore weather in science.

My dog is scared. She is shaking and whining under the bed. I sit on the rug and rub her ears. It does not seem to help. Daisy is a big floppy German Shepherd, and she is not scared of anything. Well, she is scared of one thing. Daisy hates the sound that wind makes as it hits the corner of the house.

I try to tell her it is just the air moving. I speak calmly, and I stay very quiet and still. All these things help Daisy feel safe. I describe the sounds I hear. There are four tall pine trees outside the window. The wind through their needles sounds like a quartet of violins. Air pushes past the pine needles like a bow on violin strings. A willow tree by the creek trails its long branches into the water. Air lifts the branches and whistles through leaves like a dozen flutes singing.

Daisy is shaking a little less. I think she likes my story. Her head rests on my leg, and big brown eyes look up at me through tan lashes. Her tail thumps a bit. I keep talking.

Sound starts as vibration. I thump Daisy's nametag to make it jingle. Sound hits your ears like a pendulum swinging. Little tiny hairs in your ears start to vibrate. All those branches and leaves shaking outside are vibrating the air, see? Daisy's eyes roll to the window, but she does not lift her head. Faster vibration means high pitched sound. Listen.

We close our eyes and listen to the wind. All that moving air pounds the window panes, making each piece of glass rattle like a drum. Some of the wind slips through the cracks around the window, and we hear deep mellow sounds. "Those are drums and cellos in our symphony," I tell Daisy. She is not impressed.

Windy flutes, violins, cellos, and drums play. With the moon high above and the faded rug below us, Daisy and I drift off to sleep. Some wild and blustery version of Bach's Cello Suite No.1-Prelude flies through the night. Daisy calms down as we listen to the symphony of wind.

RHOMBI'S PLAYHOUSE
STEMVESTIGATION: DESIGN

TECHNOLOGY

READ	Accidental Inventions
DISCUSS	Have you ever made a mistake that ended up working well? Like the commercial that shows chocolate accidentally dipped in peanut butter, some mistakes end in great inventions.
EXPLAIN	Companies like 3M "solve problems by applying creativity and ingenuity to make life easier." Adhesives like tape and glue are one area in which technology has changed our lives. What adhesives can we name?
MOVE	What items in the classroom / school / playground are held together at seams, sides, joints and corners? What technology is used to keep the items together? (cement, nails, tape, glue, etc.)

INVESTIGATE

Challenge: Investigate ways to connect two pieces of paper such that no wind goes through the seam. Work in teams like professionals so that your creativity is boosted by shared ideas. All teams should test at least 4 ways to connect their paper. Get creative. Innovate! What crazy ways could you come up with to keep the wind on one side of the "wall?"

Adhesive suggestions: masking tape, painter's tape, clear tape, duct tape, glue, paste and paper across the seam? What else can you test?

ELA	Keep an invention journal for the week. Draw or write about tools and technology noticed during the week!

RHOMBI'S PLAYHOUSE
STEMVESTIGATION: DESIGN

Stop the wind!

Wind sneaked in through tiny cracks.
Rhombi was cold.
How could she keep out the wind?

Cut out the square.

Cut out the square.

Accidental Inventions
Explore new ideas in technology.

1.0  POPSICLES

11 year old Frank Epperson wanted to make soda pop at home. In 1905, the popular drink could only be bought at stores and restaurants. Frank used his porch as a testing lab. He mixed powder and water to find the right taste and texture. Frank left the ingredients in a cup where he'd been working. Temperatures dropped overnight.
Frank's drink froze with the stir stick stuck in its center.

Soda Pop + Icicle = Popsicle

2.0  WAFFLE CONES

The 1904 World's Fair is the birthplace of today's popular ice cream cone. Ice cream in dishes had been around for years. A dessert seller at the Fair was doing so much business that it quickly ran out of bowls and plates. A waffle maker at the next booth was not doing much business at all. The two business owners worked together to roll up the Persian waffles and fill them with ice cream.

Ice Cream + Waffle es = Ice Cream Cone

3.0 CHOCOLATE CHIP COOKIES

The Toll House Inn's owner was named Ruth Wakefield. Ruth wanted to make some of her delicious chocolate cookies for people staying at the inn. Ruth realized that she was out of baker's chocolate. The inn keeper broke the regular sweet chocolate bar candy into little chips for the cookie dough. She expected the bits to melt into chocolate cookies. They did not. Instead, the chips stayed whole.

Dough + Chips = Chocolate Chip Cookies

4.0 STICKY NOTES

In 1968, Inventors Spencer Silver and Art Fry were researchers at 3M Laboratories. Spencer created a "low tack" sticky substance. It was unique in that it could be removed without hurting a surface. Unfortunately, no one had a use for the stuff. It went into storage. Many years later, Art Fry needed a way to keep papers in his choir's song book. He suggested a use for Spencer's sticky invention!

Low Tack Sticky + Paper = "Post-Its©"

5.0 SLINKY SPRING TOYS

Naval Engineer, Richard Jones, dropped one of the tension springs for a battleship project on the ground. It bounced from place to place all over the room. He spent about 2 years finding just the right materials and coil to turn his spring into a toy. Betty, his wife, named the new toy. "Slinky." is a Swedish word that means sleek or sinuous. In 1945, the first batch of 400 toys sold out in 90 minutes.

Coil Spring + Hard Work = Slinky ©Toys

?

What items do you use that might be combined to create a new invention?

Could you find a new use for something that you already own?

What would you invent to solve a problem or make the world a better place?

RHOMBI'S PLAYHOUSE
STEMVESTIGATION: STRUCTURES

ENGINEERING

READ	The Tiny House
DISCUSS	What do we already know about wind? How many students have watched someone use a pattern or a blueprint? Has anyone ever followed pictures to make something? What do students already know about the structures in which they live? If you unfolded your house, what do you think it would look like?
EXPLAIN	The simple, square shape of a box house means that all the parts could be unfolded to show a "net" of the building. Nets are used in mathematics to show a shape's faces. Nets are used in science to map the universe.
MOVE	Stretch using the terminology of architecture and a "Rhombi Says" format. "Rhombi says to stretch toward the sky like a skyscraper. Rhombi says to reach deep toward the basement." "Rhombi says stand tall like a column."

INVESTIGATE

Completely unfold empty cereal, cookie, pasta, and other food boxes. Trace the "nets." Refold and tape the boxes.

Cut out the "pattern" created by tracing, and fold to make paper boxes.

ELA	Explain the box pattern to a teammate.

RHOMBI'S PLAYHOUSE
STEMVESTIGATION: STRUCTURES

Trace your box's pattern.

The Tiny House
Explore structures in engineering.

The Tiny House
Explore structures in engineering.

My Grandpa and I are building a tiny house. Grandpa says his tiny house will use less electricity and water than the house he and Grandma used to share. I am helping. Most of Grandpa's tiny house will be built far away and delivered to his in big flat boxes. We will put all the different pieces together to make a whole house.

There is a lot to do before the boxes get here. First we prepare the site. The land must be "graded." All the rocks and roots are taken away. The land is made as flat as possible. We drive a little road grader back and forth, then use a laser to level the ground. Next comes the foundation.

Grandpa forms and pour concrete cubes called pier pads. We have to wait a couple of days while the concrete sets. Two 4ft x 8ft beams rest side-by-side on the concrete. This holds the floor underneath the walls and house. I use a wrench to thread nuts onto bolts. Grandpa follows behind to tighten each one. He lays the drainage pipe and wiring through floor tunnels before adding the last of the sub-floor.

Grandpa will use a special composting toilet. All waste runs into a system of barrels under the tiny house to become fertilizer for his garden. The water will come from city plumbing, but gray water (waste water from sinks) will drain into the garden also. Grandpa uses a generator to power tools for framing and roofing. Grandpa's walls were pre-made. Aunt Elizabeth comes over to prop up the walls and attach frames. We "plumb" the walls to be sure they stand up straight from the floor. I hold a "level" tool so that its bubbles are perfectly centered. We make sure windows and door are level and square. We wrap it all in sealing tape to keep out water and mold.

Next, we'll add the roof. Grandpa uses a band saw to cut the rafters that rise from walls to roof. The rafters are spaced 4 feet apart. Insulation is added to walls and roof, then layers of roofing material top off the tiny house. The house is done! Grandpa and I paint the tiny house and add shutters. We hang a sign that says, "Welcome to Grandpa's Tiny House in the Woods."

RHOMBI'S PLAYHOUSE
STEMVESTIGATION: CUBISM

READ	Cube World
DISCUSS	Look at a favorite classroom storybook. Do the images look exactly like real objects? If you were going to draw the picture, what shapes would you use?
EXPLAIN	Cubism comes from the Latin word cubus, meaning a square or cube. Artists break down objects and people to basic geometric forms. Artists also show objects from more than one angle (front, back, top) in the same painting. Picasso and Braque started the cubist movement in 1908.
MOVE	Choose a cubist painting to reenact as a class, or give each team in the classroom a different picture. Record the reenactments, and challenge other teams to choose the corresponding picture!

INVESTIGATE

Choose magazine ads to cover with geometric shapes. The goal is to end up with a recognizable image using geometric shapes in lieu of the photographic image. Determine the general shape of objects and figures in the ad. Cut 2D geometric shapes to cover objects and figures, adding hand-drawn eyes and lips and ears (chance for a sensory discussion). Add any outstanding features of objects (stripes, dots, etc.).

ELA	Describe the process of creating the "cubist" artwork.

THE ARTS

RHOMBI'S PLAYHOUSE
STEMVESTIGATION: CUBISM

Use these shapes in a drawing.

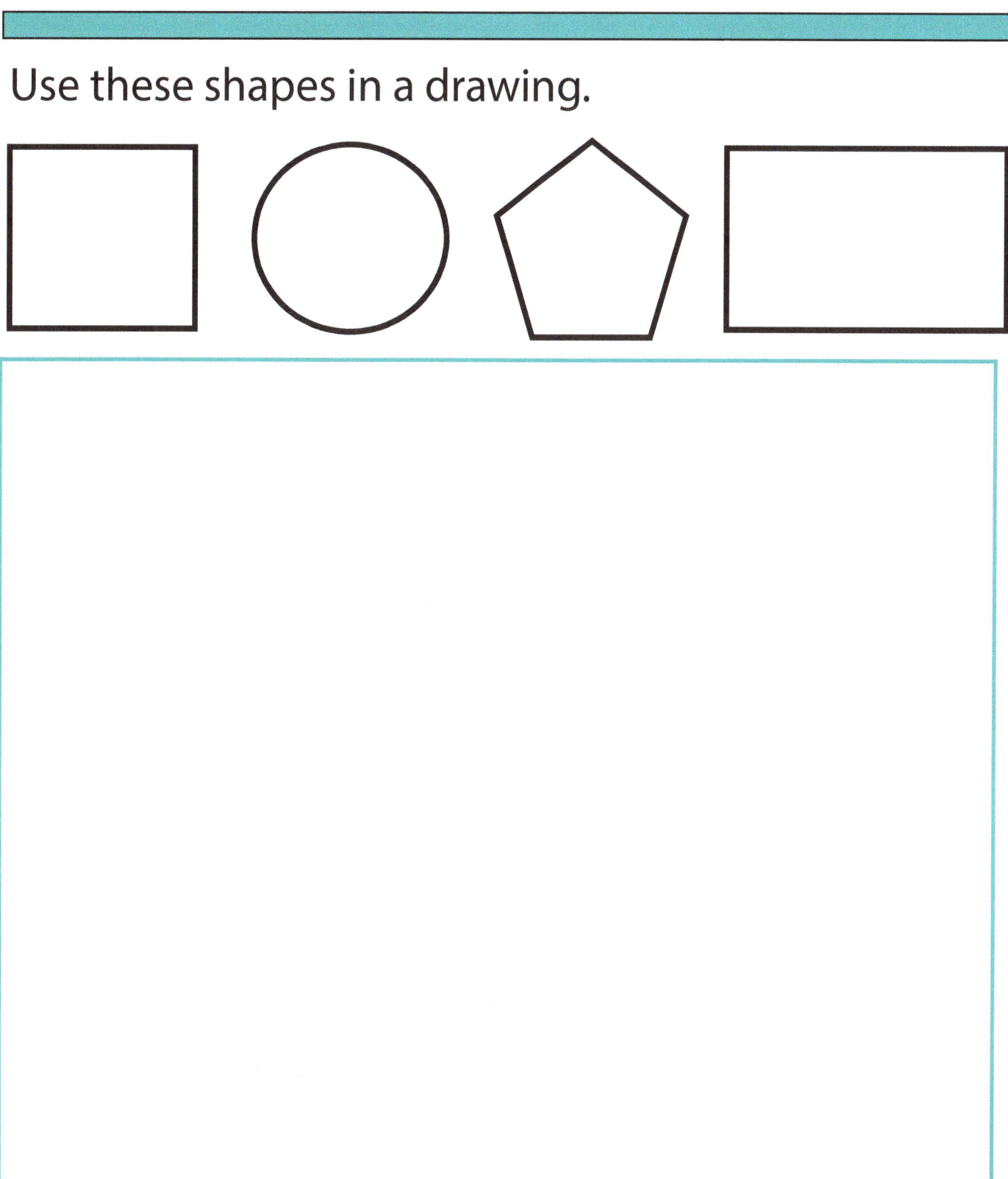

Cube World
Explore cubism in art.

Cube World
Explore cubism in art.

In the early twentieth century, artists began to look at the world as geometry in motion. Cubism began in France in 1907. Picasso and Braque started painting the world made up of cubes, spheres, cylinders, polygons, cones, and other geometric shapes.

Cubists wanted to show all the sides of an object in the same picture. The paintings looked a lot like the artist had cut the image and glued its pieces back together.

Cubist paintings show objects from more than one angle at once. Pablo Picasso and Georges Braque were first to add bright colors to cubist art. They believed that painters should not just repaint the world exactly as it exists. Instead, they wanted to show every part of the whole subject.

How does the world change when you re-imagine everything as two-dimensional and three-dimensional shapes?

RHOMBI'S PLAYHOUSE
STEMVESTIGATION: PERSPECTIVE

READ	Do You Know the Quadrilaterals
DISCUSS	What 2D shapes do you see in cubes and cylinders? If you look at a stack of cubes, can you see all the cubes that make up the stack?
EXPLAIN	Google describes *perspective* "as the art of drawing solid objects on a two-dimensional surface so as to give the right impression of their height, width, depth, and position in relation to each other when viewed from a particular point." Perspective changes as the object is turned or moved closer / farther away.
MOVE	Look at a single object from 4 or more different locations. How does the *perspective* change? Does the shape look exactly the same from every location?

INVESTIGATE

Stack blocks or math cubes for each team with 9 blocks at the base. Stack another 9 blocks at the center, and finish the stack with 9 more blocks. Even though each face of your cube full of cubes is exactly the same, the perspective makes top and side look slanted. Draw the shape. Describe the shape. Create more stacked cubes made of cubes. How many cubes are actually present in the shape? How many cubes are you able to count without removing any layers?

ELA	Explain why the cubes cannot be seen from all vantage points.

RHOMBI'S PLAYHOUSE
STEMVESTIGATION: PERSPECTIVE

How many blocks are in this stack?

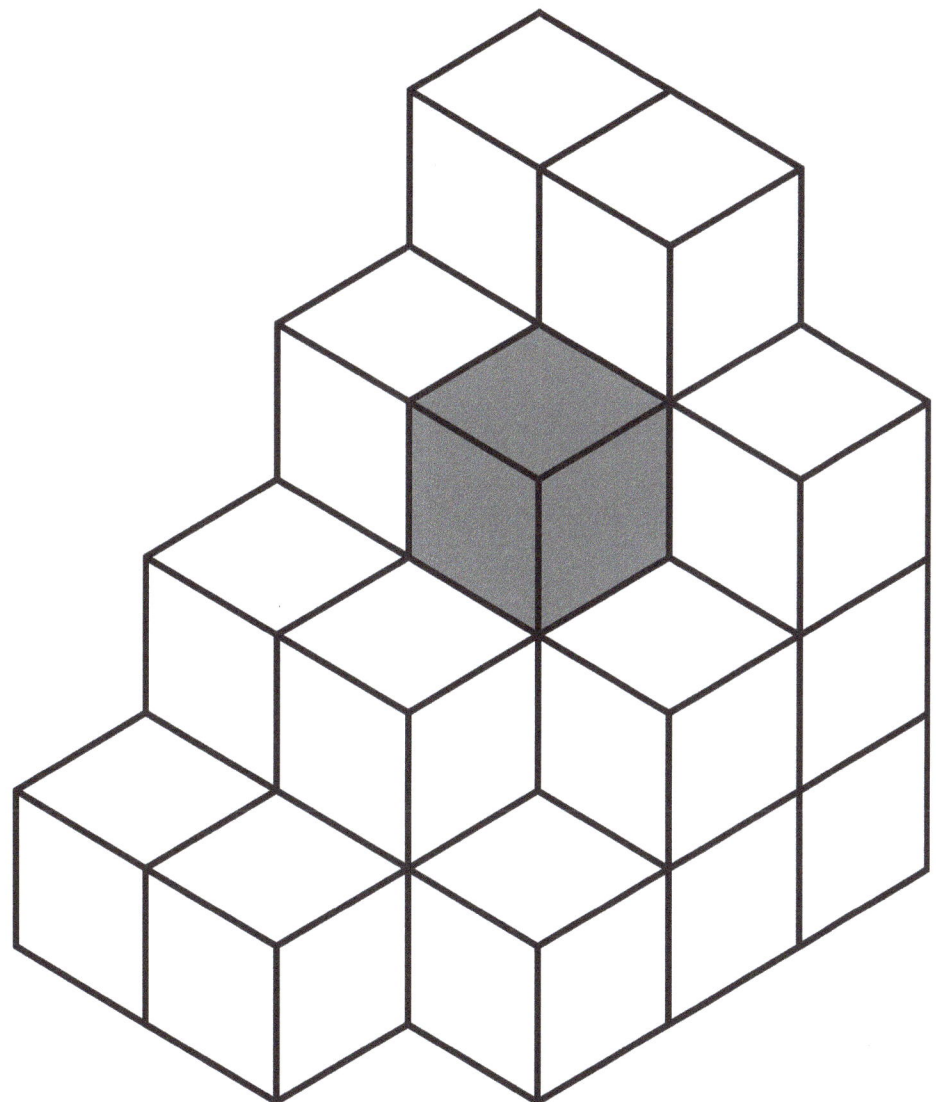

Build the stack and count.

Do You Know the Quadrilaterals
Explore quadrilaterals in mathematics.

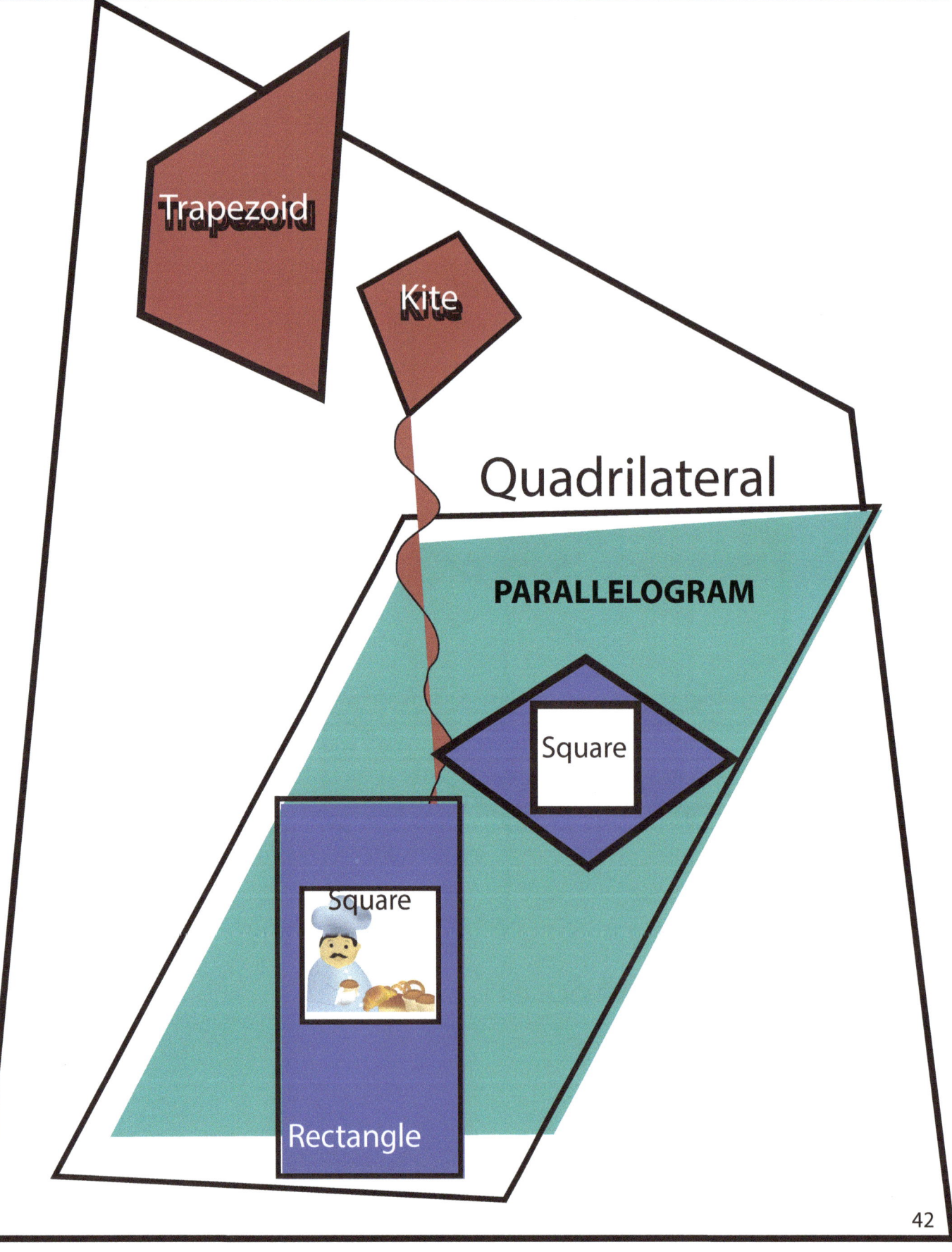

Do You Know the Quadrilaterals
Explore quadrilaterals in mathematics.

Oh do you know the quadrilateral, quadrilateral, quadrilateral?
Do you know the quadrilateral? A closed shape with 4 edges.
Oh, yes, I know the quadrilateral, quadrilateral, quadrilateral.
Oh, yes, I know the quadrilateral. 4 edges and 4 angles.

Now, do you know the trapezoid, the trapezoid, the trapezoid?
Do you know the trapezoid? It's a quadrilateral.
Oh, yes, I know the trapezoid, the trapezoid, the trapezoid.
Oh, yes I know the trapezoid. It has a pair of parallels. (a pair of LLs)

Now, do you know the parallelogram, the parallelogram, parallelogram?
Do you know the parallelogram? It's a quadrilateral.
Oh, yes, I know the parallelogram, parallelogram, parallelogram.
Oh, yes, I know the parallelogram. It has TWO pairs of parallels.

Now, do you know the rectangle, the rectangle, the rectangle?
Do you know the rectangle. It's a quadrilateral.
Oh, yes, I know the rectangle, the rectangle, the rectangle.
Oh, yes, I know the rectangle… (deep breath)
It has two sets of parallel edges where opposites edges and opposite angles are the same.

Now, do you know how to find a square, find a square, find a square?
Do you know how to find a square. It's a quadrilateral.
Oh, yes, we know how to find a square, find a square, find a square.
Oh, yes we know how to find a square. Look for all edges and all angles the same.

Now we know our quadrilaterals, quadrilaterals, quadrilaterals.
Now we know our quadrilaterals.

Trapezoid, Parallelogram, Rectangle, Square!

RHOMBI'S PLAYHOUSE
PLAY WITH SHAPES

SOCK IT TO ME

MATERIALS
Clean disposable socks for each 3D shape
cube, pyramid, cone, cylinder, rectangular prism, sphere
Black marker
stopwatch

SETUP
Place one shape in each clean sock. Number each sock.

INSTRUCT THE PLAYERS
Students teams time each other naming items in the socks. Player 1 feels the shape through the sock and names each shape in turn along with the number listed on the sock. Player 2 uses the stopwatch to record time in seconds. Player 3 double checks the numbers against the shape name on paper to make sure the student names the correct shape. Players will attempt to decrease the number of seconds required to name all objects. (Discuss why the number of seconds should go down instead of up...)

INSTRUCT THE PLAYERS
Students could easily do this activity in a station alone or with one teammate.

Rhombi's Adventures in 3D

RHOMBI'S PLAYHOUSE
UNIT 6 CHALLENGE

RHOMBI'S PLAYHOUSE: CHALLENGE

Once upon a time, there was Rhombi. Rhombi loved shapes and found them everywhere. She especially loved the squares that made up each face of a cube.

RHOMBI'S PLAYHOUSE: CHALLENGE

Rhombi's family had just moved to a new home with a yard. Rhombi had permission to build her very own playhouse. Rhombi chose a shape for her house that looked like blocks and sugar cubes.

RHOMBI'S PLAYHOUSE: CHALLENGE

Rhombi knew that she should build a sturdy playhouse to withstand the fall winds, but the last warm days of summer called. The forecast said only partly cloudy, so Rhombi was not worried about the weather.

RHOMBI'S PLAYHOUSE: CHALLENGE

Rhombi hurried to finish her building so she could get to a festival on time. She drew a sketch of her house. Rhombi decided that she could fix it before the weekend. She didn't want to miss time with her friends.

RHOMBI'S PLAYHOUSE: CHALLENGE

That night, the weather changed without warning. The air grew colder and blew wildly around the corners of Rhombi's home. Rhombi ran home.

RHOMBI'S PLAYHOUSE: CHALLENGE

Rhombi jumped in surprise as one sharp gust of wind pushed at her house. Another gust of wind pulled at the house. The walls shuddered. The house shivered. So did Rhombi.

RHOMBI'S PLAYHOUSE: CHALLENGE

She stared at all the crispy leaves as they tumbled from the trees. Then, Rhombi stared in amazement as the walls of her house also tumbled to the ground.

RHOMBI'S PLAYHOUSE: CHALLENGE

Looking at the walls piled on the cold ground, Rhombi decided she needed some help. She looked out beyond the pages of her world and asked, "Designers, can you help me create a new and stronger house?"

RHOMBI'S PLAYHOUSE
TALK ABOUT STEAM

How many 2 dimensional square and triangular faces make a 3 dimensional cube?
6

How is a 2D object different from a 3D object?
2 dimensions means the shape has length and width. 3 dimensions means the shape has length, width and height.

What is another way to say "autumn?
Fall

What is the average rainfall in your area for the month of February?
Answers will vary.

How are freezing water and thawing ice related?
Frozen water is called ice. Ice melts - or thaws- to make liquid (water).

How would rain get from the roof to the ceiling of the rooms below?
The roof is flat. The underside of the playhouse's roof is also the ceiling of Rhombi's room.

Why is Rhombi wearing rain boots and carrying an umbrella?
She wants to dress appropriately for the rainy weather.

About what is Rhombi wondering?
She wonders what to do to fix her leak.

How could Rhombi change the shape of her flat roof?
She could make it pointy or tent-like.

Why would Rhombi want to change the shape of her roof?
The water is collecting on her flat roof. If she makes a slanted roof, the water will slide off and not cause puddles or leaks through the ceiling.

What materials will you need in order to build Rhombi's playhouse roof?
1 square base and 4 equilateral triangles

RHOMBI'S ADVENTURES IN 3D
UNIT 6 DESIGN CHALLENGE: RHOMBI'S PLAYHOUSE

"Designers, can you help me create a new and stronger house?"

my idea	my plan

CUBE NET
Print this net for use in a center or during STEMVESTIGATIONS.

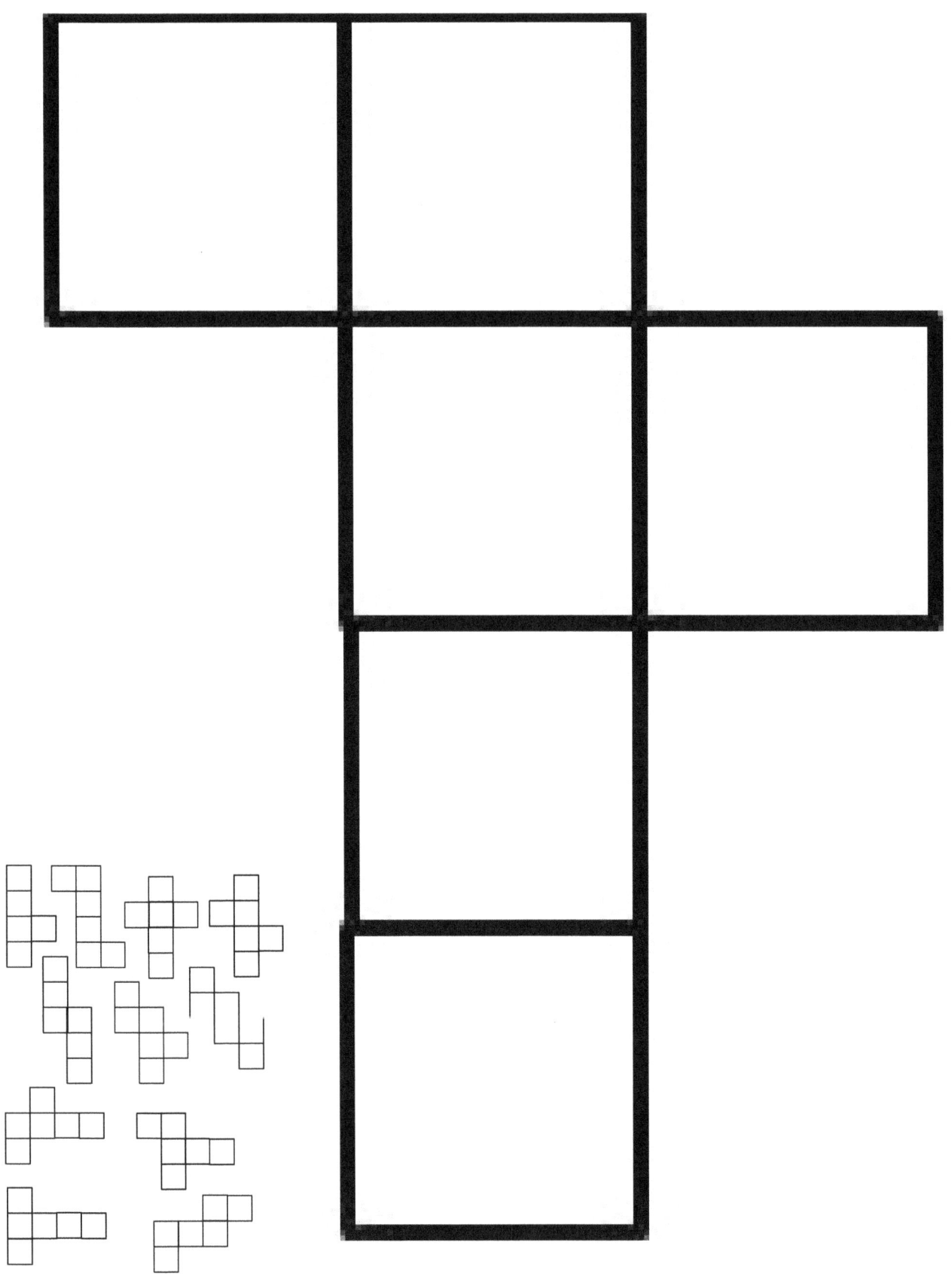

56

RHOMBI'S ADVENTURES IN 3D
RUBRIC FOR RHOMBI'S PLAYHOUSE

TASK: Students will use their knowledge of habitat, house, cube and strength to build a scale model of a playhouse that will not fall apart from the pressure of a hair dryer on high (29mph or higher).

	Content	**Organization**	**Design & Build**
1	• Is well thought out and supports the solution to the challenge or question • Reflects application of critical thinking • Has clear goal that is related to the topic • Is accurate	• Information is clearly focused in an organized and thoughtful manner • Information is constructed in a logical pattern to support the solution	• Make or draw a cube. • Uses squares to create new cubes • Neatly joins corners and edges • Remains cohesive even if blown by the hair dryer • Student adjusts concept in response to dryer test
2	• Supports the solution • Has application of critical thinking that is apparent • Has no clear goal • Has some factual errors or inconsistencies (i.e. wheels do not rotate around axle)	• Project has a focus but might stray from it at times (more concerned with form than function) • Information appears to have a pattern, but the pattern is not consistently carried out in the project	• Identifies a square but does not combine them to make a cube • Edge seams are mis-joined and uneven • Student is unable to adjust concept following dryer test
3	• Provides inconsistent information for solution • Has no apparent application of critical thinking • Has no clear goal • Has significant factual errors, misconceptions, or misinterpretations	• Content is unfocused and haphazard • Information does not support the solution to the challenge or question • Information has no apparent pattern	• Does not draw/make 2D shapes • Does not know to join squares to make a cube • Edge seams are missing or ragged • Student project falls apart and student is unable to make changes

Rhombi's Adventures in 3D

ROOF ROOF

UNIT 7

RHOMBI'S ADVENTURES IN 3D
ROOF ROOF

ROOF ROOF

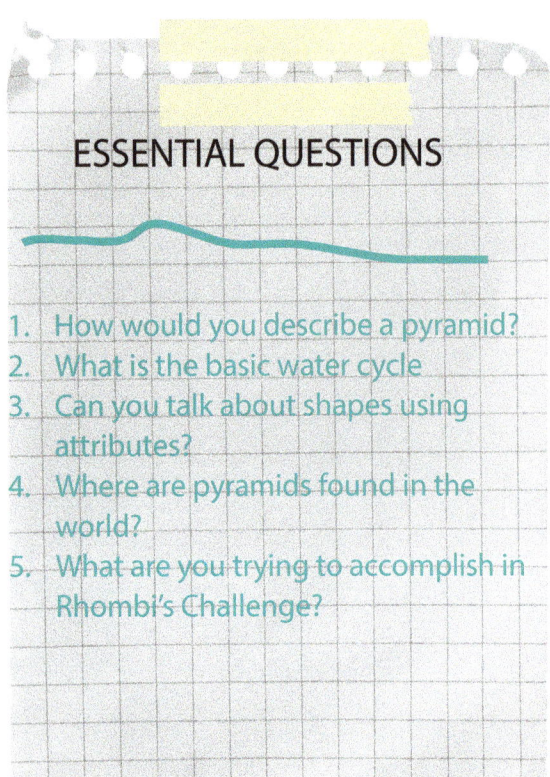

ESSENTIAL QUESTIONS

1. How would you describe a pyramid?
2. What is the basic water cycle
3. Can you talk about shapes using attributes?
4. Where are pyramids found in the world?
5. What are you trying to accomplish in Rhombi's Challenge?

Focus on triangles, squares, and pyramids while reinforcing cubes. Explore water cycles. Investigate materials with sun baked images. Decide the most efficient way to build a pyramid net. Play with height, and create a strong column. Learn more about elapsed time.

Students will apply basic measurement, understanding of number, place value, data collection, and data application in problem solving. Students will use two-dimensional shapes to create a cube and a pyramid.

Rhombi's playhouse looks great, but the roof is leaking. Students will apply new understanding of shapes, weather, materials, water cycles, and pyramids to design and construct a waterproof roof.

freezing
pass from the liquid to the solid state by loss of heat

thawing
change from a frozen solid to a liquid by gradual warming

pyramid
solid figure with a polygonal base and triangular faces that meet at a common point

waterproof
to make unaffected by water

"Halves" means two equal parts.

Circles and rectangles break into equal parts.

Breaking circles or rectangles into more parts means that the parts will be smaller.

Three objects may be put in order from longest to shortest by comparing their lengths

The length of an object may be discussed using whole numbers.

A smaller object may be used as a measurement tool when measuring longer lengths.

Larger numbers consist of tens and ones.

Organizing information helps in discussion.

ROOF ROOF
CURRICULUM CONNECTIONS

ART CONNECTIONS

PYRAMIDS
While we know that the stone for the pyramids was quarried, transported and cut from the nearby Nile, we still cannot be sure just how the massive stones were then put into place. While stone was generally reserved for tombs and temples, sun-baked mud bricks were used in the construction of Egyptian houses, palaces, fortresses, and town walls.

TECHNOLOGY CONNECTIONS

Play to Learn
Interactive Polygons
http://www.learner.org/interactives/geometry/3d_pyramids.html

HISTORICAL CONNECTIONS

PYRAMIDS are present in architecture dating to earliest recorded history. Examples include:
• Egypt- Great Pyramids of Giza
• the Americas - Pyramid of the Sun / of the Moon

According to History.com, the Americas contained "more pyramid structures than the rest of the planet combined." Mayans, Aztecs and Olmecs built pyramids for deities and burial. Latin American pyramids include the Pyramid of the Sun and the Pyramid of the Moon at Teotihuacán in central Mexico.

MATH BACKGROUND

In geometry, a SQUARE PYRAMID is a pyramid having a square base. If the apex is perpendicularly above the center of the square, it will have what is called C4v symmetry.

A Johnson solid is one of 92 strictly convex regular-faced polyhedra, but which is not uniform, i.e., not a Platonic solid, Archimedean solid, prism or antiprism. They are named by Norman Johnson who first enumerated the set in 1966.

Requires	4 triangles & 1 Square
Edges	8
Vertices	5

MINDBUGS TO NOTE

Rain comes from holes in clouds.
Rain comes from clouds sweating.
Rain occurs because we need it.
Rain falls from funnels in the clouds.
Rain occurs when clouds get scrambled and melt.
Rain occurs when clouds are shaken.
Clouds move because we move.
Clouds come from somewhere above the sky.
Empty clouds are filled by the sea.

ROOF ROOF
CURRICULUM CONNECTIONS

COMMON CORE CONNECTIONS

ELA/Literacy
RI.K.1 With prompting and support, ask and answer questions about key details in a text. W.K.1 Use a combination of drawing, dictating, and writing to compose opinion pieces in which they tell a reader the topic or the name of the book they are writing about and state an opinion or preference about the topic or book. W.K.2 Use a combination of drawing, dictating, and writing to compose informative/explanatory texts in which they name what they are writing about and supply some information about the topic.

Mathematics
MP.2 Reason abstractly and quantitatively. MP.4 Model with mathematics. K.CC.A Know number names and the count sequence. K.MD.A.1 Describe measurable attributes of objects, such as length or weight. Describe several measurable attributes of a single object. MD.A.1 Describe measurable attributes of objects, such as length or weight. Describe several measurable attributes of a single object.

NEXT GEN SCI CONNECTIONS

ETS1.A: Defining Engineering Problems
A situation that people want to change or create can be approached as a problem to be solved through engineering. Such problems may have many acceptable solutions.

1-ESS1-1,1-ESS1-2 Patterns in the natural world can be observed, used to describe phenomena, and used as evidence.

ESS2.D: Weather and Climate
Weather is the combination of sunlight, wind, snow or rain, and temperature in a particular region at a particular time. People measure these conditions to describe and record the weather and to notice patterns over time. (K-ESS2-1)

K-ESS2-1. Use and share observations of local weather conditions to describe patterns over time.

UNIT S.T.E.A.M. ACTIVITIES

Rhombi Audio Download / Video (available December 2014)

UNIT2: Roof Roof:	Science:	Water
UNIT2: Roof Roof:	Technology:	Pyramid
UNIT2: Roof Roof:	Engineering:	Height
UNIT2: Roof Roof:	Art & Music:	Materials
UNIT2: Roof Roof:	Mathematics:	Weathering
UNIT2: Roof Roof:	ELA:	Read and illustrate "Roof Roof."

ROOF ROOF
BUILD S.T.E.A.M. WITH GREAT BOOKS

SCIENCE

Suns
Franklyn M. Branley

Water Cycle (& Many Other) Subject Videos
http://www.youtube.com/user/makemegenius/videos

TECH

Water Cycle Video
Http://alturl.com/2vju3

Explore a Pyramid Interactive Experience
Http://education.nationalgeographic.com/education/kd/?ar_a=5

ENGINEERING

How
Gail Gibbons

Look
Scot Ritchie

THE ARTS

If You Lived Here: Houses of the World
by Giles Laroche

Matisse for Kids
http://www.artbma.org/flash/f_conekids.swf

MATH

Cubes
Tana Hoban

Teacher Interactive for White Board - Symmetry Flip / Rotate
http://www.teacherled.com/resources

ROOF ROOF
STEMVESTIGATION: PYRAMIDS

READ	The Case of the Disappearing Sand
DISCUSS	What do students already know about the pyramids? How wide and tall were the Egyptian pyramids? How long did they take to build? What was inside them?
EXPLAIN	Pyramids of Khufu at Giza were built thousands of years ago in Egypt. The Pyramid at Giza stands 481 feet tall (147 meters). This pyramid is a square pyramid because its base is a square and all 4 triangular faces are equilateral triangles.
MOVE	Can your class (or team) make a human pyramid? Walk the length of the pyramid's height (481 feet).

INVESTIGATE

Visit a real pyramid in the virtual world. Take a web-based field trip through the Pyramid at Giza.

Explore a Pyramid Interactive Experience
http://education.nationalgeographic.com/education/kd/?ar_a=5

Build a Pyramid Online
http://www.bbc.co.uk/history/ancient/egyptians/launch_gms_pyramid_builder.shtml

ELA	Create a comic strip to describe how the pyramid was built in your interactive experience.

ROOF ROOF
STEMVESTIGATION: WATER EROSION

Make a drip tool using the diagram.

Water Is As Water Does
Explore the states of water in science.

66

Water Is As Water Does
Explore the states of water in science.

LIQUID
Water that we drink - or spill - is called liquid. Water changes names as it changes temperature.

ICE
Water that gets very cold turns solid. Solid water is called ice.

VAPOR
When water gets warm enough, it seems to disappear but is still in the air. Water we cannot see in the air is called vapor.

WATER, WATER EVERYWHERE
How many of these phrases do you know?

Wet your whistle...
Like oil and water...
Blood is thicker than water...
Mad as a wet hen...

CELSIUS
ice 0°
vapor 100°

FAHRENHEIT
ice 32°
vapor 212°

ROOF ROOf
STEMVESTIGATION: WATER

TECHNOLOGY

READ	**Water Is as Water Does**
DISCUSS	Where can you find water? What happens to rain puddles once clouds drift away to reveal the sun?
EXPLAIN	Water on Earth is always changing. Its repeating changes make a cycle. As water goes through its cycle, it can be a solid (ice), a liquid (water), or a gas (water vapor). Ice can change to become water or water vapor. Water can change to become ice or water vapor. Water vapor can change to become ice or water.
MOVE	Be the water! Play the rain game. Sit cross-legged on the ground in a circle. The class pats hands on legs or drums the ground. Simulate a rain storm. Wind is pushed ahead of the storm. Rain falls softly, increasing in strength (slap the ground harder and harder). Thunder booms (Teacher claps hands together.) Rain decreases in strength until the sound dies away altogether.

INVESTIGATE
Choose 4-6 different surfaces to test: include something porous like paper and something nonporous like plastic. Give each team of students responsibility for one surface. Place 5 drops of water on each surface. Measure the width of the puddle. Place the test surfaces in a sunny window or under a lamp. Make a coordinate graph showing the hour of the day vs. width of each puddle.

| ELA | Each hour of the day, take digital photos of the surfaces. Make a slide handout in PowerPoint. |

ROOF ROOF
STEMVESTIGATION: PYRAMIDS

Make a pile of sand.
Draw the pile of sand each day this week.

Monday	Tuesday	Wednesday

Thursday	Friday

The Case of the Disappearing Sand
Explore the data collection in technology.

Read the charts with data collected by Rhombi and Pi.
What story does the data tell?
What happened to the sand sculpture?

DAY 1 THE PYRAMID BASE IS 3 FEET SQUARE 4 FEET HIGH SUNNY WIND 5 MILE PER HOUR
DAY 2 BASE IS ALMOST 4 FEET SQUARE 3 FEET HIGH SUNNY WIND 10 MILES PER HOUR
DAY 3 BASE IS 2 FEET ON THE LONGEST SIDE BUT NOT ALL SIDES ARE EVEN. 2 FEET HIGH ON ONE SIDE ALMOST 3 FEET HIGH ON THE OTHER SIDE (NO LONGER SYMMETRICAL) RAIN LAST NIGHT WINDS 25 MILES PER HOUR
DAY 4 THE BASE IS NOT SQUARE 1 FOOT HIGH ON ONE SIDE ONE SIDE HAS LOST A LOT OF SAND RAIN WINDS GUSTING TO 4 MILES PER HOUR
DAY 5 MOSTLY FLAT AND NOT MUCH SAND LEFT

The Case of the Disappearing Sand
Explore data collection in technology.

The school bell rang. Kids tumbled out of the building. "What are you doing this weekend," Rhombi asked her friend Pi. Pi didn't have any plans. Neither did Rhombi. Maybe they would go for a swim or meet at the park? Rhombi and Pi saw a bright orange flier stuck to the school bulletin board.

"Sand Sculpture Contest Saturday at 9AM"

The two friends looked at the sign. They looked at each other. They knew exactly what they'd be doing this weekend.

Rhombi and Pi hauled buckets, spades, small shovels, and lots of cardboard boxes to use as sand molds. The team got name tags and were assigned a 5ft by 5ft area of sand. They sketched out a quick design. Rhombi used clay to mold a model of the sculpture. When both friends were happy with the design, they started to build. They dug. They piled. They formed and pressed. Finally, the sand started to look like a pyramid. Small camels milled around at the base, and sand tents held tiny people. About 4PM, Rhombi and Pi stepped back to look at their finished sculpture. It looked great!

Rhombi and Pi watched as judges made notes at each team's area. There were some amazing designs: seagulls flew over bridges; tsunamis crashed on beaches; castles held dragons, and a couple of robots held hands. The judges called all the contestants over to announce winners. Rhombi and Pi got bright green participation ribbons. They cheered as 1st, 2nd and 3rd place winners were given medals. Then everyone headed home, tired and sandy but proud.

Rhombi and Pi passed the beach each day on the way to school. They noticed something interesting. The sculptures were changing shape. Each day, their pyramid design changed just a bit. The two decided to make a timeline that showed the changes every day for a week. Each day for a week, they made a fast sketch of the sculpture and jotted down notes to keep a good record. They also tracked rain with a measuring cup system and wind with an anemometer. The weather detectives could solve this case. What was making the sand disappear?

STEAMStart copyright 2020 by Jeannie Ruiz All Rights Reserved

ROOF ROOF
STEMVESTIGATION: PYRAMIDS

READ	Circle, Triangle, Square
DISCUSS	What familiar buildings have pointed rooftops? How tall are the "pyramids" or "cones?" What other names can you think of for the tall upper part of a building? Why might buildings need pointed tops? (wind, rain)
EXPLAIN	Some of the tops of buildings are cones with circle bases and pointed top. Some buildings are topped with a pyramid shape (square base and triangular faces). Some have spires.
MOVE	Students lie down head to feet to make a human ruler. How many students would need to stand on each other's shoulders to touch the tallest part of the Chrysler Building? According to CNN: 25 Great Skyscrapers: Icons of Construction, the spire rises 186 feet above the main building.

INVESTIGATE
How tall can a top (apex) rise before the building topples the pyramid? What steps can be taken to keep the building stable no matter how tall the spire rises? Challenge: Build the tallest pyramid spire possible before the building topples. Measure the height of the building for each attempt. Measure the height of the spire for each attempt. Graph building height vs. spire height.

ELA	Each hour of the day, take digital photos of the surfaces. Make a slide handout in Power Point.

ENGINEERING

ROOF ROOF
STEMVESTIGATION: ROOF HEIGHTS

How tall can you make a tower before it falls?

Color the number of blocks you stacked.

Circle, Triangle, Square
Explore rooftops in engineering.

What is the shape of your rooftop?

Circle, Triangle, Square
Explore rooftops in engineering.

Circle, Triangle, Square
How many rooftops touch the air?

Round like the top of a castle turret.
Round like the top of a tower.
Round like the clock that's hand strikes each hour.

Slanted like the ridge of a Swiss chalet.
Angled like the top of a tepee.
Pointed like a spire in the city.

Flat like a hospital's helicopter pad.
Flat like the tar on a high rise.
Flat like the planetarium, mirrors pointed at the sky.

Circle, Triangle, Square
How many rooftops touched the air?

ROOF ROOF
STEMVESTIGATION: MATERIALS

READ	I Love Homes page
DISCUSS	Have you ever made mud pies? Clay figures? How long does your creation take to dry? Do all clays dry?
EXPLAIN	Around the world, bricks are made of local materials. From clay to soil and dirt, people use the tools at hand to build homes. Clay or dirt alone will crumble. Adding grass, straw or string keeps the brick together. Sculptors also use clay mixtures to create works of art. Women of the Southwestern Native American tribes normally created pots, and the men decorated them with designs like simple geometric patterns.
MOVE	Look for clay, straw and soils around the building that might be used to make tiny pyramid.

INVESTIGATE

Make the Mixture: Bricks may be made when clay soil is mixed with straw. Mix soil with water until it becomes quite thick. Add the straw. How much straw should be used? How much water? What is the best mixture to make a clay shape that dries without crumbling? You might use candy or cookie molds, or just form figures with fingers. Make a scale model of a pyramid once you've determined the best mixture of clay, dirt, straw... Leave the structure out in full sun for 2 days to bake the clay. Decorate!

ELA	Watch BBC instructors build bricks! http://www.bbc.co.uk/learningzone/clips/how-are-bricks-made/13460.html

THE ARTS

ROOF ROOF
STEMVESTIGATION: MATERIALS

Circle the outcome. Does the paper fall apart?

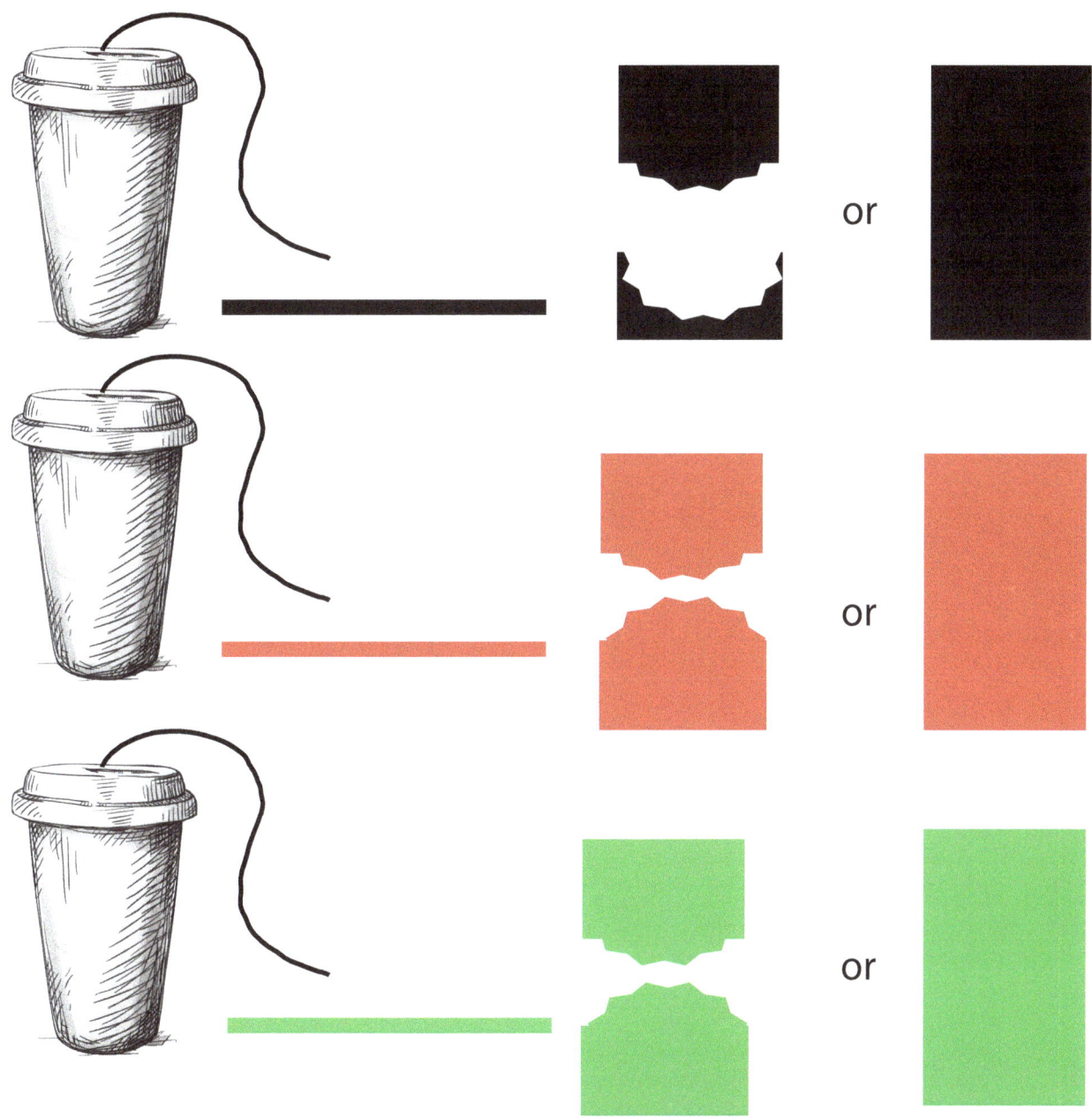

I Love Homes
Explore houses in art.

I Love Homes
Explore houses in art.

I love homes.
Tall homes… Small homes…
Round homes and square
Homes in big buildings with hundreds of stairs.

I love homes:
Blue homes… Yellow homes…
Red homes and white
Homes with black shutters to close out the night.

I love homes.
Prairie homes… Hobbit holes…
Towers and castles
Yurts in the grassland and tents in the desert.

I love homes.
Stilt homes… Ranch homes…
Igloos and trailers
Maybe I'll live on a boat like a sailor.

I love homes.
Cottage homes… Apartment homes…
Cabins or condos
Tiny houses or duplexes lined up in neat rows.

I love everyone's home.

What kind of home do you love?

ROOF ROOF
STEMVESTIGATION: WEATHERING

MATHEMATICS

READ	The Camel and the Pyramid
DISCUSS	Why do people use pyramids? If rain is the main reason, why would pyramids be used in Egypt? Wind is also a factor: wind has less surface to push when the building narrows to a point. Less wind, less erosion.
EXPLAIN	The Egyptian pyramids are more than 3000 years old. The structures are not as sharp as they once were. Wind and rain break down materials over time.
MOVE	Use a sheet, rod, square base, and heavy-duty binder clips to make a pyramid tent. Teamwork is required to build the structure! Make a hole in the center of your square base (cardboard works well). Insert the tall rod, and have one student hold it steady. Drape a sheet over the rod. Secure sides / corners with binders clipped to the base.

INVESTIGATE

Make a mixed clay pyramid. Once baked in the sun, measure and record the length of sides and height of the pyramid. Students that are too young to measure with a ruler should place a piece of paper or cardboard behind the structure, and trace its original outline. Place the pyramid in a tray. Leave the pyramids in a sunny area protected from wind. Every 3rd day (maintain a steady pattern), redraw or measure the pyramid.

Repeat the same process, but place the pyramid in a windy place protected from the sun. Look at mountains in Europe vs. those in American Appalachia. US mountains are significantly older. They've been worn down and rounded...

ELA	Make a class geography board, and add stickers or pins to chart locations found in stories through the year.

ROOF ROOF
STEMVESTIGATION: EROSION

Make a drip tool.

The Camel and the Pyramid
Explore patterns in mathematics.

The Camel and the Pyramid
Explore patterns in mathematics.

There is a camel that lives in the far away sand of the Egyptian desert. His tan fur is stiff and rough. It has to keep out the tiny sand crystals as they crash through air during windstorms. He can close his nostrils and a third eyelid to keep out the terrible sharp sand. He has a hump that stores fat. The camel can live for days without food as his body uses the hump's fat to stay alive. Camels are tough.

The camel's caretaker is 14 years old. The young camel's guide wears a kafiyeh, a head wrap that does the same job as his camel's rough hide and closed nostrils. The scarf wound around the boy's head keeps sand and sun away from the boy's face. Wind blows lightly at 9 or 10 miles per hour through most days, but windstorms from the desert can be harsh. The wind may whip around the pyramids at 150 miles per hour. This kind of wind is dangerous to boys, camels and pyramids.

The mahout knows that the pyramids will last a long time, though. As he guides tourists from all over world, he tells the history of these pyramids. He tells the families and backpacking teenagers that The Great Pyramid was made of horizontal rows of bricks. About 2,500,000 bricks were stacked like blocks. Each brick weighed around 2.5 tons. That means each brick weighed more than a car or a full-grown male giraffe. Tourists, happy with his stories, gave him tips called baksheesh. The boy earned a lot of baksheesh to help care for his camel and help feed his family.

On this day in March, the winds are blowing. There are no tourists to hear his tales or ride his camel. The young mahout and the 800 pound camel head home to strong stables. They need to escape the sand and wind. As they plod away toward Cairo, the mahout looks over his shoulder at The Great Pyramid. It would not blow away in the storms, even though little bits of its faces were eroded by the constant wind. Thousands of years would pass before the wind and water slowly brought the pyramids to the ground.

The camel lowered his third eyelid, closed his nostrils, and headed home toward fresh water and good food and a dry place away from the itching, scouring winds.

ROOF ROOF
PLAY WITH SHAPES

LEARN TO JUGGLE

MATERIALS
Soft round balls (3 per student or enough for a single station)

SETUP
Head outside, or clear desks for room to juggle.

INSTRUCT THE PLAYERS
Ancient Egyptian art shows images of Egyptians juggling balls! You can learn to do this, too.

INSTRUCT THE PLAYERS
Students could easily do this activity in a station alone or with one teammate. You'll need to master one level of juggling before moving to the next level.

1. Toss one ball from one hand to the other. Practice until you can toss the ball as high as your eyes.
2. Hold one ball in your weaker hand. Hold a second ball in your stronger hand. Toss the strong hand ball, and -just before you catch it in your weak hand- move the weak hand ball to the strong hand (switch).
3. Repeat above but toss the second ball up and over instead of just switching hands.
4. When you are comfortable with 2 balls, add a third.

source http://learnhowtojuggle.info/the-basics/

Rhombi's Adventures in 3D

ROOF ROOF

UNIT 7 CHALLENGE

ROOF ROOF CHALLENGE

Once upon a time there was Rhombi. She had a strong playhouse shaped like a cube. Autumn winds had not been able to push it down. Now that it was winter, she loved to hang out with a book and a warm blanket.

ROOF ROOF CHALLENGE

While the playhouse was snug and strong, the roof might need a redesign. Water had caused problems. Rain had made pools of water that sat on the roof. Snow and ice had frozen and thawed to make more puddles.

ROOF ROOF CHALLENGE

Drip. Drip. Drip. Freezing water and thawing ice had caused a hole in the roof! She worried that her books might be ruined. Rhombi imagined a waterfall in her house.

ROOF ROOF CHALLENGE

Rhombi's dad used a ladder to check out the playhouse roof. He looked over the flat roof full of puddles.

He said, "You have a real hole up there. I think you need to redesign the roof."

ROOF ROOF CHALLENGE

What kind of roof would she need? Rhombi stood outside with her umbrella and rain boots. She looked at her wet bird/squirrel feeder's steep roof. If she had a roof like that, water could not pool and make puddles.

ROOF ROOF CHALLENGE

She looked out beyond the pages of her world and asked, "Designers, can you help me create a waterproof roof for my playhouse?"

ROOF ROOF
TALK ABOUT STEAM

How many 2 dimensional square and triangular faces make a 3 dimensional cube? **6**

How is a 2D object different from a 3D object?
2 dimensions means the shape has length and width. 3 dimensions means the shape has length, width and height.

What is another way to say "autumn? **Fall**

What is the average rainfall in your area for the month of February? **Answers will vary.**

How are freezing water and thawing ice related?
Frozen water is called ice. Ice melts - or thaws- to make liquid (water).

How would rain get from the roof to the ceiling of the rooms below?
The roof is flat. The underside of the playhouse's roof is also the ceiling of Rhombi's room.

Why is Rhombi wearing rain boots and carrying an umbrella?
She wants to dress appropriately for the rainy weather.

About what is Rhombi wondering?
She wonders what to do to fix her leak.

How could Rhombi change the shape of her flat roof?
She could make it pointy or tent-like.

Why would Rhombi want to change the shape of her roof?
The water is collecting on her flat roof. If she makes a slanted roof, the water will slide off and not cause puddles or leaks through the ceiling.

What materials will you need in order to build Rhombi's playhouse roof?
1 square base and 4 equilateral triangles

RHOMBI'S ADVENTURES IN 3D
UNIT 7 DESIGN CHALLENGE: ROOF ROOF

"Designers, can you help me create a waterproof roof for my playhouse?"

my idea	my plan

RHOMBI'S ADVENTURES IN 3D
UNIT 7 DESIGN RUBRIC: ROOF ROOF

TASK — Students will use their knowledge of habitat, house, pyramids and strength to build a scale model of a roof that will allow water to fall from the roof without puddling.

	Content	Organization	Design & Build
1	- Is well thought out and supports the solution to the challenge or question - Reflects application of critical thinking - Has clear goal that is related to the topic - Is accurate	- Information is clearly focused in an organized and thoughtful manner - Information is constructed in a logical pattern to support the solution	- Makes or draw a pyramid - Uses triangle and square to make a pyramid - Corners and edges are neatly joined - Remains cohesive even if "rained" on... - Student adjusts concept in response to "rain" test
2	- Supports the solution - Has application of critical thinking that is apparent - Has no clear goal - Has some factual errors or inconsistencies (i.e. wheels do not rotate around axle)	- Project has a focus but might stray from it at times (more concerned with form than function) - Information appears to have a pattern, but the pattern is not consistently carried out in the project	- Identifies triangle and square but does not combine them to make a pyramid - Edge seams are mis-joined and uneven - Student is unable to adjust concept following "rain" test
3	- Provides inconsistent information for solution - Has no apparent application of critical thinking - Has no clear goal - Has significant factual errors, misconceptions, or misinterpretations	- Content is unfocused and haphazard - Information does not support the solution to the challenge or question - Information has no apparent pattern	- Does not draw/make a pyramid - Does not join triangles and squares to make pyramid - Edge seams are missing or ragged - Student project falls apart and student is unable to make changes

PYRAMID NET
Print this net for use in a center or during STEMVESTIGATIONS.

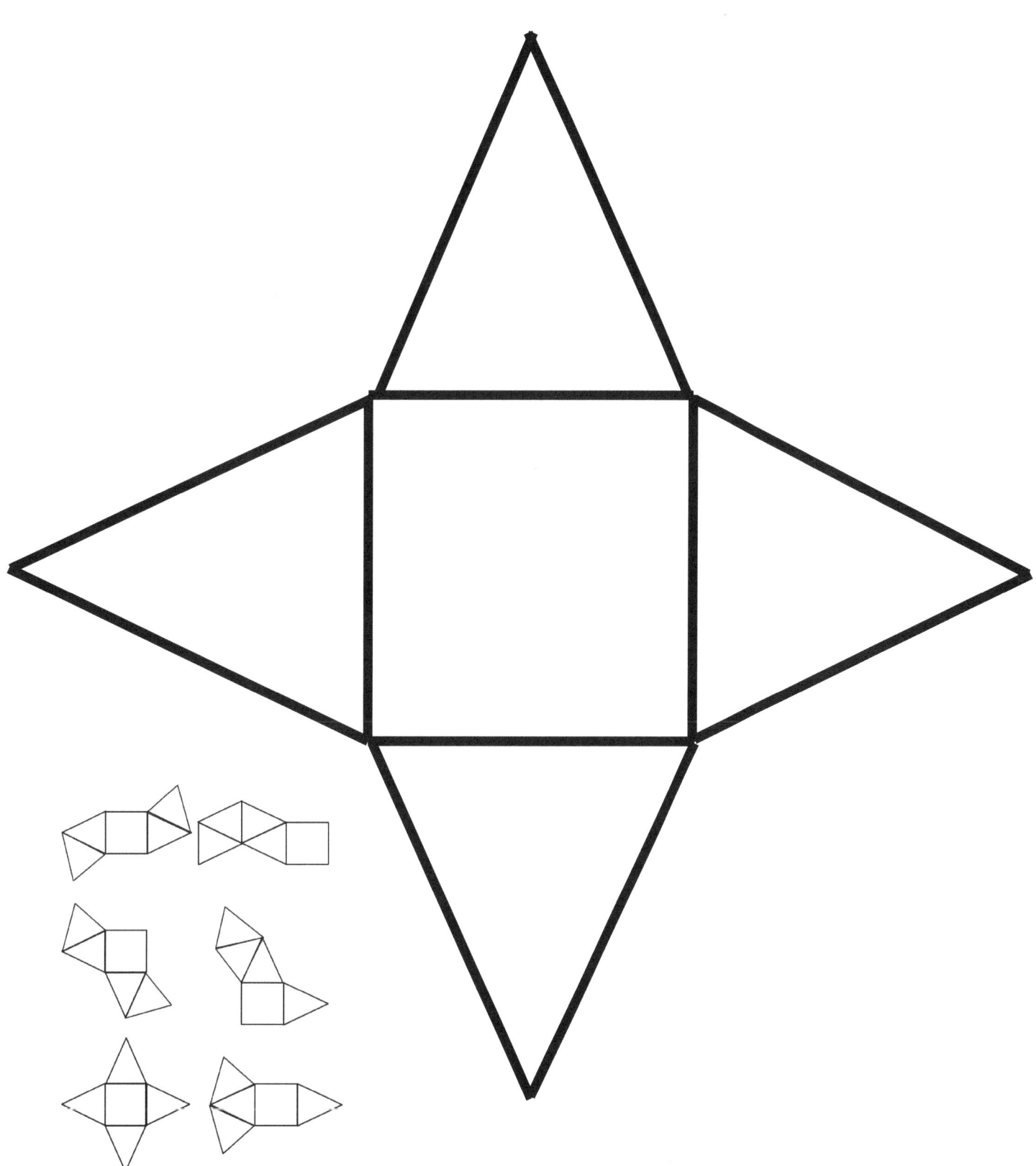

Rhombi's Adventures in 3D

PET FINDS A HOME

UNIT 8

RHOMBI'S ADVENTURES IN 3D
PET FINDS A HOME

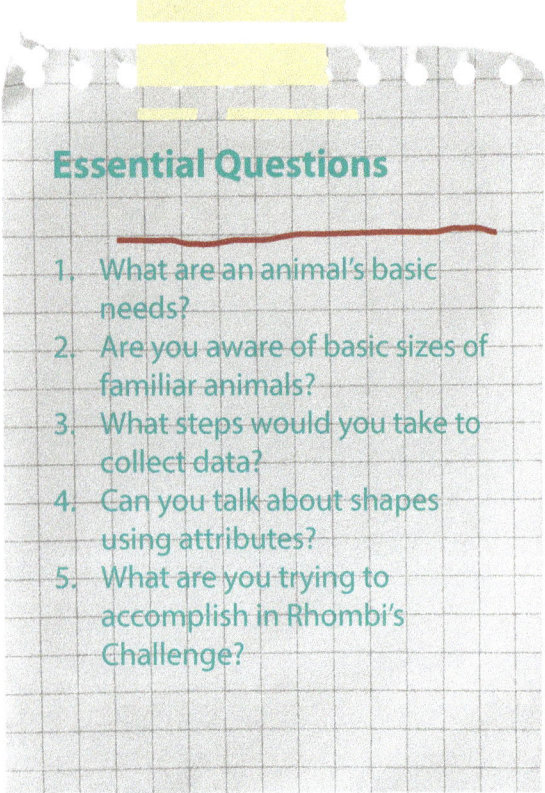

Essential Questions

1. What are an animal's basic needs?
2. Are you aware of basic sizes of familiar animals?
3. What steps would you take to collect data?
4. Can you talk about shapes using attributes?
5. What are you trying to accomplish in Rhombi's Challenge?

ENDURING UNDERSTANDING

Circles and rectangles break into equal parts.

Three objects may be put in order from longest to shortest by comparing their lengths.

The length of an object may be discussed using whole numbers.

A smaller object may be used as a measurement tool when measuring longer lengths.

Larger numbers consist of tens and ones.

Organizing information helps in discussion.

Collecting, representing and analyzing data is an important part of decision-making.

PET FINDS A HOME

Focus on combining two-dimensional shapes to create three-dimensional shapes for a purpose. Investigate basic needs. Discover the natural habitat of the South American guinea pig. Develop a strong foundation. Play with patterns, and look at size and scale.

Reinforce for students that may have missed the mark in Units 1 and 2. Accelerate for students ready to take on a challenge. Combine cubes and pyramids.

Rhombi welcomes a new pet into her playhouse. She and Pet trip all over each other, and Rhombi decides that her tiny playhouse needs a special room for Pet. Students design and build the scale model for a pet habitat (pet of their choice).

basic needs
something all living things must have to survive such as nutrients, water, sunlight, space, and air

habitats
natural home of an animal, plant, or organism

living things
something that needs food, water, and air to live and grow

PET FINDS A HOME
CURRICULUM CONNECTIONS

Rhombi's Pet needs a place of her own.

ART CONNECTIONS

REALISM TO CUBISM
From the realistic watercolors of Albrecht Dürer to the cubist oils of Franz Marc, animals have been a popular topic of art through the ages. Work with the school media specialist and art teacher to incorporate drawing, painting and sculpture on the subject of animals in art. Visit this great URL for more information and examples: www.artcyclopedia.com/subjects/Animals.html

TECHNOLOGY CONNECTIONS

Art Projects for Kids: Animal Habitat Sketches
http://www.artprojectsforkids.org

New York Zoo's Website: Create Your Wild Self
http://buildyourwildself.com/

HISTORICAL CONNECTIONS

ANIMALS AS PETS
While scientists are still unable to pinpoint the exact location, they can say with a fair amount of certainty that the "gray wolf" was the first animal to become man's best friend. When humans began settling for longer periods of time, they also started dumping refuse in piles outside the camps. Wild wolves were drawn in and eventually became familiar with humans. Through many generations, humans bred the animals for guard duty and hunting. Dogs have many jobs now, including that of "best friend."

MATH BACKGROUND

In geometry, height is defined as the measurement from an object's base to its top.

In a pyramid, height would be measured from the center of the base to the pinnacle of the pyramid.

In a cube, the height should be measured from base of one edge to the top of the same edge (to avoid accidentally slanting to one side or another).

MINDBUGS TO NOTE

Developing a sense of size and scale is crucial to problem solving reasonableness.

SCALEVILLE: BUILD A WORLD OF NUMBERS
Start a wall chart that builds through the year. Like a "word wall," the chart adds data as it is recognized in various subjects. Reading about hippos should prompt a card with the hippo's height placed at the correct height on your SCALEVILLE wall. Add each kid's height. Add the heights of plants, animals, vehicles, and anything else that makes sense.

Building a sense of size and scale will help students with problem solving.

PET FINDS A HOME
CURRICULUM CONNECTIONS

COMMON CORE CONNECTIONS

ELA/Literacy
RI.K.1 With prompting and support, ask and answer questions about key details in a text. W.K.1 Use a combination of drawing, dictating, and writing to compose opinion pieces in which they tell a reader the topic or the name of the book they are writing about and state an opinion or preference about the topic or book. W.K.2 Use a combination of drawing, dictating, and writing to compose informative/explanatory texts in which they name what they are writing about and supply some information about the topic.

Mathematics
MP.2 Reason abstractly and quantitatively. MP.4 Model with mathematics. K.CC.A Know number names and the count sequence. K.MD.A.1 Describe measurable attributes of objects, such as length or weight. Describe several measurable attributes of a single object. MD.A.1 Describe measurable attributes of objects, such as length or weight. Describe several measurable attributes of a single object.

NEXT GEN SCI CONNECTIONS

ETS1.A: Defining Engineering Problems
A situation that people want to change or create can be approached as a problem to be solved through engineering. Such problems may have many acceptable solutions.

1-ESS1-1,1-ESS1-2 Patterns in the natural world can be observed, used to describe phenomena, and used as evidence.

ESS2.D: Weather and Climate
Weather is the combination of sunlight, wind, snow or rain, and temperature in a particular region at a particular time. People measure these conditions to describe and record the weather and to notice patterns over time. (K-ESS2-1)

K-ESS2-1. Use and share observations of local weather conditions to describe patterns over time.

UNIT S.T.E.A.M. ACTIVITIES

Rhombi Audio Download / Video (available December 2014).

UNIT 3:	Pet Finds a Home:	Habitats
UNIT 3:	Pet Finds a Home:	Attributes
UNIT 3:	Pet Finds a Home:	Shapes
UNIT 3:	Pet Finds a Home:	Patterns
UNIT 3:	Pet Finds a Home:	Scale and Size
UNIT 3:	Pet Finds a Home:	ELA: Read and illustrate "Pet Finds a Home."

PET FINDS A HOME
BUILD S.T.E.A.M. WITH GREAT BOOKS

SCIENCE

A Drop In The Ocean: The Story Of Water
by Jacqui Bailey

Classroom Library:
favorite animals
classroom pets
habitats
needs of living things

TECH

Joe the Gentle Giant and the Badger's Home
by Wayne Stripling

Classroom Library:
animal classification
adaptation

ENGINEERING

Wheels at Work
by Dan Zerbrowski

Classroom Library:
towers
civil engineers

THE ARTS

Pattern Fish
by Judy Harris

Classroom Library:
M.C. Escher
tessellations
patterns in nature

MATH

Life-Size Farm:
Teruyuki Komiya

Classroom Library:
measurement, small animals, medium animals, large animals

PET FINDS A HOME
STEMVESTIGATION: HABITAT

READ	**Meghan Finds a Purrl**
DISCUSS	Look at the book covers. What makes each animal unique? What things do the animals have in common?
EXPLAIN	Animals need air, food, water, and a safe place to raise young. Pets also need to breath, eat, drink and feel safe in their homes. As the caretaker, you are responsible for providing all these things needed for living things to survive and stay healthy.
MOVE	March through the classroom or playground chanting, "food, water, air and home." As students return to seats, each student should repeat the list to an instructor.

INVESTIGATE

Students bring stuffed animals or toy pets to class. Students will use construction paper, scissors, tape, art materials and cardboard scraps to design and build habitats for their pets. Focus on shapes. Present the animal's habitat in terms of geometry during a show-and-tell using appropriate vocabulary. The bowl is a circle. The cage is a cube. The door is a rectangle. The hamster's food is a pile of spheres. Etc.

ELA	Add an additional technology element with an interview booth. Hang a sheet on the wall. Students make their presentations to a video camera.

PET FINDS A HOME
STEMVESTIGATION: HABITATS

Make an animal show-and-tell.

Pick an animal.

Draw a home.

Draw foods that the animal will eat.

Meghan Finds a Purrl
Explore perspective in science.

One rainy day, Meghan and Davis were splashing around in mud puddles. They saw a kitten curled up in a cold, sad little ball by the curb. It looked miserable and wet. Meghan tried to pick it up, but the kitten hissed and tried to scratch her arm. Davis pulled off his sweatshirt. They wrapped the angry kitten in Davis's warm sweatshirt. Thank goodness the kitten calmed and began to purr. It fell asleep.

Meghan washed and brushed the little black ball of fur. She sat in front of the heater's vent with the kitten on her lap. It was so soft!

The veterinarian talked to the Davis and Meghan about healthy food and habits. He made them promise to have the kitten registered and vaccinated. That would keep the kitten from getting diseases later in life. The veterinarian even helped them pick kitten food. They picked a soft food for nutrients and a crunchy food to keep the kitten's teeth healthy.

The kids named their new friend Purrl and got her a bright orange collar. Meghan gave the kitten toys and a bed. She watched Purrl grow taller, and she noticed one day that her pet's eyes had turned green. Purrl got patch of white fur grew on her chest and white "socks" on her paws. Purrl was growing quickly.

Purrl slept on the Meghan's pillow every night and begged for breakfast scraps in the morning.

As Meghan held her purring cat, she whispered, "They all lived happily ever after.
Good night, Purrl."

Purrl Finds a Home
Explore perspective in science.

Once upon a time, there was kitten. She was lonely and cold. Rain fell on her fur and dripped in her sad blue eyes. She huddled at the curb trying to get out of the wind.

She saw two giants looking down at her. Cold wet dripped down from giant shiny bodies. Their giant hands reached for her, and she hissed. The kitten's fur stood up in tufts, and her tail stuck straight out behind her body. She hopped, trying to look big and scary. The world got dark and a little bit warmer. Still, the kitten shook and hissed and meowed. She struggled to get free of the dark dry thing, but it was warm and dry. The kitten calmed and began to purr. She was dry. She was warm. She slept.

When she woke, the giants petted her fur. She ate some yummy brown paste and crunchy cookie food. She drank clear, cool water. She curled up by a hot wind and fell asleep again.

The kitten woke to find a thing around her neck. She clawed at the thing. It made jingly noises. She chased it in a circle. The giants laughed. She hissed, but she didn't really mean it…

The kitten washed her fur every day. She grew as tall as the giant's knees. The giants got names. The one with long head fur that smelled like vanilla was Meghan. The one with short head fur that smelled like grass was Davis. Grassy David visited a lot.

Purrl slept most of the day waiting in the window for sun to touch her paws. That's when the big stinky thing spit Meghan out onto the sidewalk every afternoon. Then she and her Meghan played until bedtime.

Purrl purred as the two best friends settled in for the night.

PET FINDS A HOME
STEMVESTIGATION: ATTRIBUTES

TECHNOLOGY

READ	Rhombi Lived in a Zoo
DISCUSS	What do we know about animals and their habits? Do fish like to sunbathe on a hot rock? Do dogs make nests in trees to protect their eggs? An animal's body gives clues to where the animal lives and what it eats.
EXPLAIN	Animal attributes match food sources and homes. Example: Animals that swim need fins and gills. Animals with wings can live in a tree or rooftop. Furred animals are able to survive colder weather. Scaled animals are protected on hot surfaces.
MOVE	Call out a habitat, and have students become animals that can survive and thrive in that environment.

INVESTIGATE

The New York Zoo has a great website on which students may mix-and-match to create their own "wild selves." Have students create a list of what they want to eat, where they want to live and what they want to be able to do. Then visit the website and create a "self" that fits the description they've created.

New York Zoo's Website
http://buildyourwildself.com/

ELA	Design a "bio" card for the "self" created at the zoo's site. How will the card be organized to show the following information; name, height, weight, description, food source, habitat, and favorite activities.

PET FINDS A HOME
STEMVESTIGATION: ATTRIBUTES

Which animal is bigger? Cut out the animals and paste them in order from tallest to shortest.

Rhombi Lived in a Zoo
Explore habitats and data collection in technology.

Rhombi Lived in a Zoo
Explore habitats and data collection in technology.

There once was a Rhombi who lived in a zoo. She had 26 rescues and life was a hoot. She fed them all dinner with fruits, veggies, protein, dairy, and bread. Rhombi checked on each one before going to bed.

Anna the ape was tucked in her nest.
Bella the boa liked sun warmed rocks best.
Calvin the colt needed plenty of hay.
Dobbie the deer slept someplace new every day.
Ella the Eastern Gorilla slept under the stars.
Fred the ferret curled up in her arms.
Gina Giraffe slept standing up tall.
Hannah the hedgehog curled up very small.
Iggy the inchworm had a leafy bower.
Jasper the jaguar was just waking up at this hour.
Kara the kitten had a basket and blankie.
Her friend Latrice Lemur climbed a tree with her hankie.
Mable the mastiff took up a whole bed.
Nonny the newt liked her fishbowl instead.
Odetta Otter burrowed into a hole.
Polly the Pig gets a wallow or away she will bolt.
Qbert the quail must sleep alone,
But Rory the rabbit wants friends in his home.
Selena the sloth bunks down on a tree limb.
Tito Terrapin carries his bed around with him.
Una Umbrella Bird has a blanket in her 48 inch cage.
Vixen the fox prefers a crevice with sage.
Wally the wren chooses a nest up high.
XRay the tetra sleeps surrounded by bubbles all the time.
Yuri the yak flops in the dirt on his back.
Zen Zebra curls up in the hay.

There once was a Rhombi who lived in a zoo. She had 26 rescues and life was a hoot. Rhombi makes notes in her file each day. She updates the spreadsheet to show eating habits and play. Who is not sleeping? Who bumped its head? Which animals might want shade or sun instead? Rhombi finished her work, shut down her computer and crawled into bed. "I'll do it all again tomorrow," she happily said.

PET FINDS A HOME
STEMVESTIGATION: SHAPES AND STRENGTH

ENGINEERING

READ	It All Adds Up
DISCUSS	Structures are built from the bottom up, a lot like the 3D piece in the story. Every strong building begins with a solid, sturdy foundation.
EXPLAIN	Structures can hold more weight when the load is shared with other shapes. The load is shared when a flat "floor" is secured above the support shapes. The floor spreads out weight, so there isn't too much weight on any one cup.
MOVE	Everybody, drop for some push-ups. It's the body forming a bridge!

INVESTIGATE

Place one cardboard square on the floor, and a single cup upside down on the square. Place the second square on top. Add weight until the cup bends or breaks. How much weight did the one cup support? Try two cups and retest. Try three cups and retest. Add your data to a class graph with cups on the x axis and weight on the y axis.

Using the data collected, decide how many cups you think you'll need to hold a student's weight. Once your structure is built, ask a teacher to help you step onto the "floor." How few cups will hold your weight?

ELA	Tell the story from the cup's point of view!

ROOM FOR PET
STEMVESTIGATION: SHAPES

Draw the rest of the building.

It All Adds Up
Explore 3D printing in engineering.

Peg watched her dad add layer after layer to the towering party cake. Cake, icing, cake, icing, cake, icing and on and on until the cake was ready to serve. Dad carefully slid the cake off its platform. He removed the little stilts that held the cake still as he iced the layers. Finally, Dad and Poly delivered the finished dessert to the museum for a party.

Dad dropped Peg off at school on the way back to the bakery. Peg ran to her classroom. She was excited about designing for the cool little machines. When Peg arrived at her station in the classroom, she sat down at her wiggly table. She stuffed a piece of folded paper under one of the legs and tried not to get annoyed with the wobbling.

It All Adds Up
Explore 3D printing in engineering.

After fixing the table, Peg found a note. It read, "Innovators look for new ways to solve problems. How will you improve the classroom using our new technology? Sketch. Model. Design. Create!"

Peg's teacher introduced the new printer technology, and explained that these machines were not like a 2D paper and ink printer. 3D printers used computer aided digital (CAD) files to make computer models. 3D printing turns the CAD computer models into real physical things. They melt materials into thin layers on a surface, adding layer on layer until the full object is made. It sounded sort of like making a cake with Dad. He sketched and designed and layered. Leaning on her table, Peg's elbow slipped off as the table wobbled again. "Ugh, this table is so distracting," fumed Peg.

Suddenly, she had an idea. Peg got down on her stomach to take a closer look at the table leg. It was shorter than the other legs, and that made the whole table wobbly. Peg grabbed her logbook and started drawing. She remembered the little stilts her dad used to even out cakes for icing. She measured the length of the other three legs, and drew a sort of stilt for the table. Peg got modeling dough to make a 3D model of her idea. She asked for help in using the scanner and watched as the CAD file was created to turn her idea into a program the printer could read. Her scanned clay model was digitally sliced into thousands of layers in the CAD program. Each of those layers would add up to make her stilt.

Peg held her breath as the first layer of her "Peg Leg" design was melted onto the 3D printer's surface. Slowly, layer by layer, the object was made. At the very end of the day, Peg pulled her "Peg Leg" from the printer's shelf. She broke off the little filaments that held it to the shelf, and hurried to her table. Kneeling, Peg slipped the 3D printed plastic piece under the short table leg. No more wobbly tables! As she headed out to meet her Dad at the car, Peg wondered how she could make her "Peg Leg" even better. Maybe there was a way to change the height? Maybe she could design different shapes for square or round legs. The ideas were endless, and they just kept adding up…

PET FINDS A HOME
STEMVESTIGATION: PATTERNS

READ	M.C. Escher's World
DISCUSS	What is a pattern? Describe some patterns you may have seen around our room. What patterns are found outside our classroom (pine cone, fences, pavement).
EXPLAIN	From a wall calendar to ceiling tiles, patterns are found all around us. Some patterns cover a space by mixing and matching shapes. Demonstrate patterns from polygons using an online interactive "Tessellation Creator." http://illuminations.nctm.org/Activity.aspx?id=3533
MOVE	Take a walk, and look for patterns. Challenge students to locate at least 10 patterns around the school.

INVESTIGATE
Practice with tessellation patterns. Print a pattern on card stock. Kids trace the pattern anywhere on a sheet of paper. Fit the shapes against each other to cover the whole page. Decorate so that no shapes that touch have the same color. Challenge kids to create a pattern that covers an entire sheet of paper. Stick to squares and rectangles to make the exercise simpler. Incorporate your own lessons on color, size and attributes.

ELA	Make a word wall with color, size and texture focus.

THE ARTS

ROOM FOR PET
STEMVESTIGATION: PATTERNS

Design a tessellation. Your pattern should repeat all over the page. What shape will you use?

M.C. Escher's World
Explore tessellations in art.

"I don't grow up. In me is the small child of my early days" -M.C. Escher

http://www.mcescher.com/about/biography/
http://euler.slu.edu/escher/index.php/Regular_Division_of_the_Plane_Drawings
http://www.csun.edu/~lmp99402/Math_Art/Tessellations/tessellations.html

M.C. Escher's World
Explore tessellations in art.

Escher was a well-known graphic artist. He composed visual material for printing. Escher lived from 1898 to 1972. Like DaVinci and Michelangelo, M.C. Escher was a leftie. He used his left-hand to write and do most activities. Also like these great artists, Escher made a lot of art!

In his lifetime, the graphic artist made 448 lithographs, woodcuts and wood engravings and over 2000 drawings and sketches. Escher also illustrated books, designed tapestries, created postage stamps and painted murals.

After visiting a Moorish castle called Alhambra, Escher became interested in the tiled floors and ceilings. He started drawing special patterns called tiling or tessellations. Escher filled five notebooks with 137 of these tessellation drawings.

Tessellations (or tiling) are created when a shape is repeated over and over again. Each shape fits right up against the next to fill an entire surface. The shapes do not overlap or leave space between them. He described ways to make a tessellation pattern.

You can make pattern pictures like Escher's tessellations using polygons. When you've practiced filling an entire page with geometric shapes, you will be ready to try animals, plants and trees.

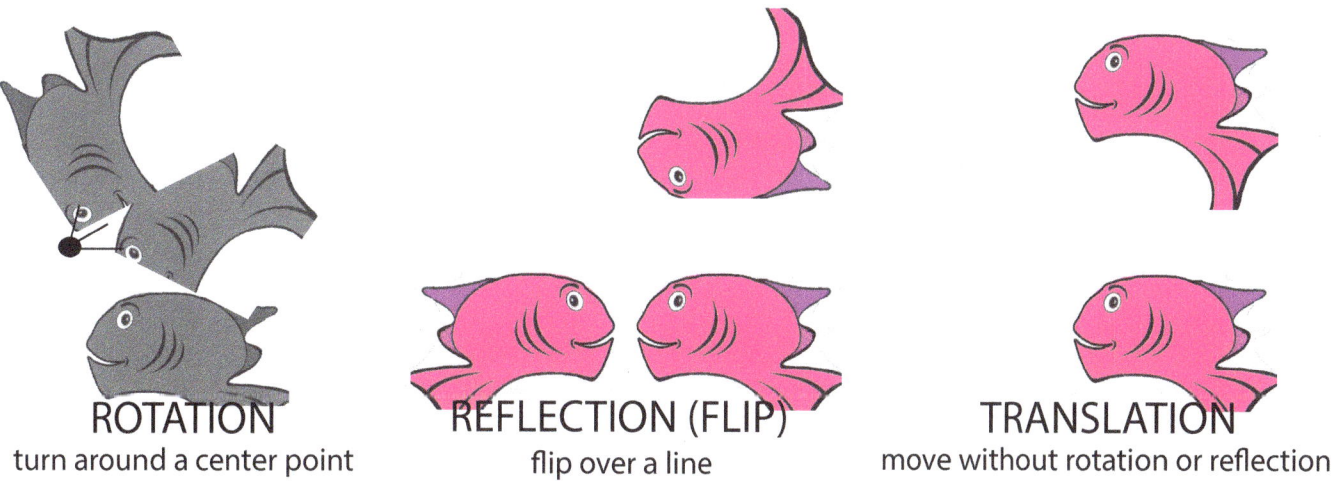

ROTATION
turn around a center point

REFLECTION (FLIP)
flip over a line

TRANSLATION
move without rotation or reflection

PET FINDS A HOME
STEMVESTIGATION: SIZE AND SCALE

READ	How Big Is a Guinea Pig
DISCUSS	Animals come in all shapes and sizes. Let's name some dogs that fit these words; tiny, small, average, big, large, and giant. Tiny = Teacup Poodle Small = Jack Russell Average = Golden Retriever Big = German Shepherd Large = Mastiff or Sheepdog Giant = Great Dane
EXPLAIN	Sometimes agreeing on what sizes fit in a category is tricky. An animal one person considers tiny might just be small for someone else. Create a system of measurement so that all descriptions agree.
MOVE	Call out a descriptor. Students have 10 seconds to locate and stand next to an item that matches the size. For example, call out, "small." Students stand next to their "small" items. Each student must name and defend choices. Play in teams or as a whole class.

INVESTIGATE

Small groups work together to devise a set of measurements that fits each adjective for size. Use measurements, pictures, or animals as the standard. Create a chart from drawings or cut-and-paste images.

Size Word	From	To	or
tiny	ant	hamster	1/2 to 5 inches
small	guinea pig	rabbit	51/2 in to 1 foot
average	cat	boxer dog	13 inches to etc....
big	bobcat	deer	**or**
large	great dane	pony	80 lbs to 200 lbs
giant	horse	elephant	201 lbs to 1 ton
ELA	Use similes to describe pets. Ex: The dog is as big as a chair.		

PET FINDS A HOME
STEMVESTIGATION: SIZE AND SCALE

Create a scale to help you talk about size. Will you use animals, plants, or objects as your guide?

SIZE	FROM	TO
TINY		
LITTLE		
SMALL		
AVERAGE		
BIG		
LARGE		
GIANT		

How Big Is a Guinea Pig

Explore size in mathematics.

How Big Is a Guinea Pig
Explore size in mathematics.

Food, Water, Air, Home
Guinea pigs never live alone.

In groups of 5 or 10, wild guinea pigs use ½ inch claws to dig burrows in the ground for sleeping and keeping their tiny pups safe. Sometimes, the little community will use burrows left by other animals or bed down in rocky crevices. The guinea pigs work as a team, screaming to warn each other of predators or other danger.

Like rabbits, guinea pigs have incisor teeth that keep growing through their lives. Eating coarse grasses helps wear down the incisors. Premolars and molars crush and grind food. Also like rabbits, guinea pigs are actually rodents. They are not pigs at all!

Guinea pigs are mammals. Babies, called pups, drink mama sow's milk. Guinea pigs have fur, and they are born live and ready to wiggle. Pups grow quickly. From a few ounces to 2 or 3 pounds in less than a year, a full grown adult might be 10 to 14 inches long. That's about the length of lined paper or legal paper.

Guinea pigs live in their communities eating grass, sleeping through the day in safe burrows, and breathing the clear mountain air of South America from Columbia to Argentina.

Now that you know how guinea pig rodents live in the wild, how would you make a guinea pig happy in a classroom habitat?

PET FINDS A HOME
PLAY WITH SHAPES

HOOP AND HOLLER

MATERIALS
hula hoops

SETUP
Place a hula hoop per team on the ground about 30 feet from the kids.

INSTRUCT THE PLAYERS
You'll need one hoop per team.
Divide the class into several small teams.
The first player of each team runs to their team's hula hoop which is placed about 30 feet in front of each team.

The team does five jumps in and out of the hoop and returns to the group.
The first team to finish the relay wins.

INSTRUCT THE PLAYERS
Form teams of 4.
Line your team up in front of a hula hoop (or jump rope).

Rhombi's Adventures in 3D
PET FINDS A HOME
UNIT 8 CHALLENGE

PET FINDS A HOME CHALLENGE

Once upon a time there was Rhombi. She had a sturdy home shaped like a cube. A pyramid roof protected her house from the sun, wind, and rain. Sunny spring days had finally arrived.

PET FINDS A HOME CHALLENGE

Rhombi took a walk. She saw all kinds of pets with their people. There were puppies. There was a bunny. She saw a guy walking with a lizard on his shoulder. Two kids with their cat made up her mind. She wanted a kitten.

PET FINDS A HOME CHALLENGE

She picked a scruffy little cat. Pet got a bath, a new collar, and a pillow on Rhombi's bed. Each night the tiny kitten curled up right under Rhombi's chin. The kitten had plenty of room to run and play.

PET FINDS A HOME CHALLENGE

The trouble started in the first week of May. The kitten was all grown up, and she needed a lot more room to move. Rhombi found Pet staring out the window at the full moon. The cat mewed and scratched at the window pane.

PET FINDS A HOME CHALLENGE

Rhombi started to worry about the cat. Pet was such a happy kitten, Rhombi wondered what was making her so grumpy now? Halfway through May, Rhombi caught Pet tearing paper in the bathroom.

PET FINDS A HOME CHALLENGE

At the end of May, Rhombi threw up her hands when she found Pet stomping through the paint. Little paw prints covered the floor.

PET FINDS A HOME CHALLENGE

Rhombi was worried about Pet. The cat seemed unhappy. Rhombi tripped over Pet two times in one morning. Rhombi made a decision about the cat.

PET FINDS A HOME CHALLENGE

Pet needed her own place to play. Rhombi looked at lots of different types of pet houses, but she could not make a decision. Rhombi looked out from the pages of her world and asked, "Designers, will you create a home for Pet?"

PET FINDS A HOME
TALK ABOUT STEAM

How many square faces does a pyramid have? **1**

How many triangular faces does a pyramid have? **4**

What is the total number of faces on a pyramid? **5**

Make a list of all the pets you can remember. **Answers will vary.**

Use the list to make a tally sheet counting students' pets.

Turn the tally sheet into a class bar graph.

What kind of animal did Rhombi choose?
The class decides on Rhombi's animal.

At what temperature do you feel chilly? **Most people feel chilly at temperatures lower than 79°**

Why would Rhombi keep Pet inside? **Pet would get too cold outside.**

What food would Rhombi eat for breakfast? What does Pet eat? **Answers will vary.**

Make a list of breakfast foods and tally food eaten by students.

What animals have a tail? **Answers will vary.**

What sounds would a grumbling pet make? **Ex: whine, chirp or meow, etc.**

Why might Pet scoot across the room? **She might be chasing toys.**

What area of floor would be "tiny" for you? Measure and test. **Ex: A 4ft x 4ft room is tiny.**

What do you think Rhombi has decided? **She will need a room for pet.**

Why does Rhombi need a special room for Pet? **She and Pet need more room to move.**

What environment is needed for a puppy? a cat? a snake? a bird? a frog? a butterfly?

What materials will you need in order to build Rhombi's pet's house? **Answers will vary depending on the choice of pet.**

RHOMBI'S ADVENTURES IN 3D
UNIT 8 CHALLENGE: PET FINDS A HOME

my idea

my plan

RHOMBI'S ADVENTURES IN 3D
UNIT 3 RUBRIC: PET FINDS A HOME

Students will use their knowledge of habitat, house, pyramids and strength to build a scale model of a habitat for a Pet.

	Content	Organization	Design & Build
1	• Is well thought out and supports the solution to the challenge or question • Reflects application of critical thinking • Has clear goal and reasons for designing this habitat for the pet • Is accurate	• Information is clearly focused in an organized and thoughtful manner • Information is constructed in a logical pattern to support the solution	• Uses 2D shapes to create 3D shapes • Neatly joins corners and edges • Constructs habitat that reflects the needs of a chosen pet
2	• Supports the solution • Has application of critical thinking that is apparent • Has no clear goal • Has some factual errors or inconsistencies (hamsters do not live in fish bowls full of water) • Student is unable to adjust concept when original idea is questions	• Project has a focus but might stray from it at times (more concerned with form than function) • Information appears to have a pattern, but the pattern is not consistently carried out in the project	• Identifies a square but does not combine them to make a cube • Edge seams are mis-joined and uneven • Constructs habitat that does not reflects the needs of a chosen pet
3	• Provides inconsistent information for solution • Has no apparent application of critical thinking • Has no clear goal • Has significant factual errors, misconceptions, or misinterpretations	• Content is unfocused and haphazard • Information does not support the solution to the challenge or question • Information has no apparent pattern	• Does not draw/make 2D shapes • Does not know to join squares to make a cube • Edge seams are missing or ragged • Student project falls apart and student is unable to make changes

Rhombi's Adventures in 3D

HAPPY BIRTHDAY, PET

UNIT 9

RHOMBI'S ADVENTURES IN 3D
PET FINDS A HOME

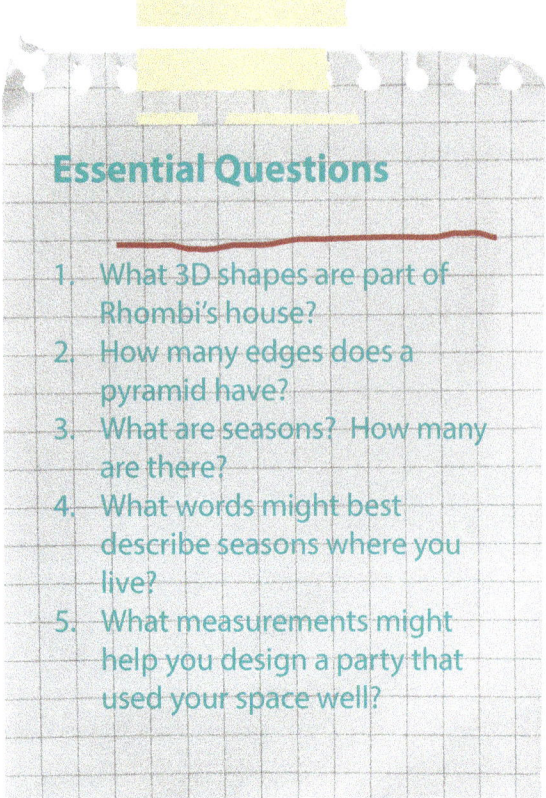

Essential Questions

1. What 3D shapes are part of Rhombi's house?
2. How many edges does a pyramid have?
3. What are seasons? How many are there?
4. What words might best describe seasons where you live?
5. What measurements might help you design a party that used your space well?

ENDURING UNDERSTANDING

Weather is what we describe in local terms. Climate describes weather patterns over a large area and over long times.

Patterns in the natural world can be observed and used to solve problems or understand what we see in nature.

A smaller object may be used as a measurement tool when measuring longer lengths.

Larger numbers consist of tens and ones.

Organizing information helps in discussion.

Collecting, representing and analyzing data is an important part of decision-making.

PET'S BIRTHDAY

Focus on combining two-dimensional shapes to create three-dimensional shapes for a purpose. Investigate basic needs of animals and people. How much space is required for animals of different sizes to be comfortable and healthy?

Use this unit to make sure all students can use basic measurement (tandard or non-standard) to help solve this unit's challenges. Accelerate for students reaabout dy to take on a challenge. Combine cubes and pyramids.

Rhombi must decide how to accomodate eight guests at a party. Students design and build the scale model for Pet's party.

basic needs
something all living things must have to survive such as nutrients, water, sunlight, space, and air

habitats
natural home of an animal, plant, or organism

living things
something that needs food, water, and air to live and grow

size and scale a
ways we compare living things to the world in which we live as humans.

HAPPY BIRTHDAY, PET
CURRICULUM CONNECTIONS

Pet is having a party and needs a space to celebrate.

ART CONNECTIONS

According to WiseGeek.org, mosaic is an art form that involves fitting small pieces of rock, shell, tile, or glass called tesserae together to create a pattern that may be abstract or representational. Students at Eisenhower Junior High School in Taylorsville, Utah hold the World Record for Largest Post-It Mosaic (38,400 pieces of paper).

Visit HistoryWorld.net for examples of mosaics.

TECHNOLOGY CONNECTIONS

Visit QR Stuff to generate codes for use in the classroom. To share the code, paste it into a Word Doc or print directly from the page.

HISTORICAL CONNECTIONS

"A column is a vertical pillar that is used to support the structure of a building. In classic architecture, a column rests on a base and is mounted by a lid, called a capital." The Greeks used columns to build open-air structures.

The three styles of Greek column are Doric, Ionic, and Corinthian. Examples of Greek columns may be seen in the following buildings:
Rome, Italy's Pantheon and Colosseum;
Egypt's Temple of Amon
and USA's Lincoln Memorial.

MATH BACKGROUND

"A GOOD unit of measurement," writes Robert Crease, "must satisfy three conditions." It has to be easy to relate to, match the things it is meant to measure in scale (no point using inches to describe geographical distances) and be stable. The yardstick did not always meet those criteria.

Its history is sketchy, but most accounts agree that the yardstick was coined by King Edward I (reigned 1272-1307) who declared, "three feet make one yard." The yard was based on the length of a pendulum that took one second to complete its swing.

MINDBUGS TO NOTE

ASSESS NUMERACY
Hold several objects in your right hand. Show the student.
Assess how far along the spectrum your student has progressed:
- Physically touches each item.
- Mentally touches each item.
- Recognizes the number of objects.

Physical contact required: issue manipulatives for all counting and problem solving.
Mental contact required: ready for activity books and handouts.
Recognition: ready for addition and subtraction.
*Test all students, including those identified as accelerated.

HAPPY BIRTHDAY, PET
CURRICULUM CONNECTIONS

COMMON CORE CONNECTIONS

ELA/Literacy
RI.K.1 With prompting and support, ask and answer questions about key details in a text. W.K.1 Use a combination of drawing, dictating, and writing to compose opinion pieces in which they tell a reader the topic or the name of the book they are writing about and state an opinion or preference about the topic or book. W.K.2 Use a combination of drawing, dictating, and writing to compose informative/explanatory texts in which they name what they are writing about and supply some information about the topic.

Mathematics
MP.2 Reason abstractly and quantitatively. MP.4 Model with mathematics. K.CC.A Know number names and the count sequence. K.MD.A.1 Describe measurable attributes of objects, such as length or weight. Describe several measurable attributes of a single object. MD.A.1 Describe measurable attributes of objects, such as length or weight. Describe several measurable attributes of a single object.

NEXT GEN SCI CONNECTIONS

ETS1.A: Defining Engineering Problems
A situation that people want to change or create can be approached as a problem to be solved through engineering. Such problems may have many acceptable solutions.

1-ESS1-1,1-ESS1-2 Patterns in the natural world can be observed, used to describe phenomena, and used as evidence.

ESS2.D: Weather and Climate
Weather is the combination of sunlight, wind, snow or rain, and temperature in a particular region at a particular time. People measure these conditions to describe and record the weather and to notice patterns over time. (K-ESS2-1)

K-ESS2-1. Use and share observations of local weather conditions to describe patterns over time.

UNIT S.T.E.A.M. ACTIVITIES

Rhombi Audio Download / Video (available December 2014)

Unit 4: Happy Birthday, Pet:	Science:	Tools
Unit 4: Happy Birthday, Pet:	Technology:	Digital Scales
Unit 4: Happy Birthday, Pet:	Engineering:	Perspective Drawing
Unit 4: Happy Birthday, Pet:	Art & Music:	Mosaics
Unit 4: Happy Birthday, Pet:	Mathematics:	Length

Unit 4: Happy Birthday, Pet: ELA: Read and illustrate "Happy Birthday, Pet."

HAPPY BIRTHDAY, PET
BUILD STEAM WITH GREAT BOOKS

SCIENCE

Butterfly Birthday
Harriet Ziefert

Classroom Library
Pets
Nutrition

TECH

Who Made This Cake?
by Chihiro Nakagawa

Classroom Library
Measurement
Scales
Construction
Weight and Mass

ENGINEERING

If You Lived Here: Houses of the World
Giles Laroche

Classroom Library
Bridges
Travel
Animal Migration

THE ARTS

Birthdays Around the World Library
by Mary D. Lankford

Classroom Library
Mosaics
Eric Carle Books

MATH

The Big Birthday Surprise: Life Lessons with Junior
Dave Ramsey

Classroom Library
Measurement, How String Is Made

HAPPY BIRTHDAY, PET
STEMVESTIGATIONS: TOOLS

READ	That's Bananas
DISCUSS	When is your birthday? Let's mark the calendar. How do you celebrate? Do you have any traditions? What do you eat? Do you let your pets celebrate with you? Do they eat the same foods as people?
EXPLAIN	Dogs cannot eat chocolate, and cats cannot drink milk. Most pets get sick when eating avocado, including dogs, cats, horses, birds and rodents. Also skip the nuts, raisins and plums. So, what would you serve to pets that attend your party?
MOVE	Mash. Mash. Mash to make some homemade treats!

INVESTIGATION

Prepare for the activity by slicing at least 2 bananas (and any other fruit you want to include) per student and freezing. To cut chaos, go ahead and freeze the fruit in individual baggies. Double bag for safety's sake. Provide cups, saucers, cardboard pieces, and other cast-off materials for constructing the "smooshers."

You are challenged to make a smooshy mooshy "ice dreamy" treat safe for pets to eat. Using materials provided, create a food "smoosher" that will mash the frozen fruit before it thaws. You'll end up with a creamy healthy treat that's safe for everyone (unless you are allergic to bananas).

When done, taste test the delicious snack!

ELA	Create a visual guide to making the smooshy treat.

HAPPY BIRTHDAY, PET
STEMVESTIGATION: DESSERT

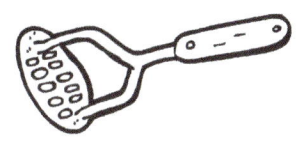

Create a tool for making mashed fruit desserts.

Choose a shape.

Draw a masher.

Draw what you will need.

That's Bananas
Explore foods in science.

Bananas are a giant herb. They do not grow on trees.

A bunch of bananas is called a hand.

The word banana comes from an Arab word "banan" meaning find bananas.

Ecuador is the world's leading exporter of bananas.

An individual banana is called a finger.

That's Bananas
Explore foods in science.

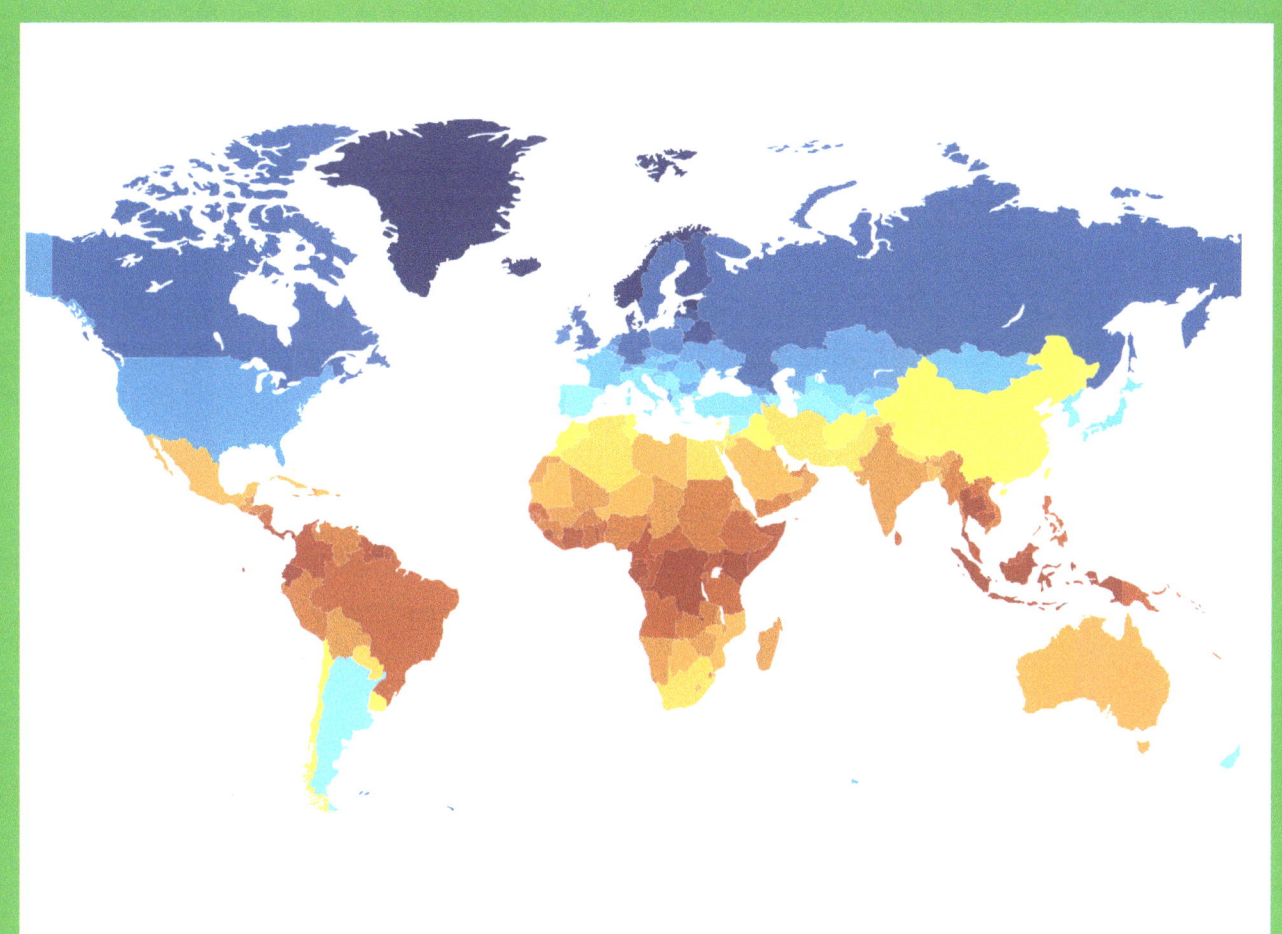

Most of the bananas you buy are grown within 20 degrees on either side of the Earth's equator in Costa Rica, Guatemala, Honduras, and Panama

HAPPY BIRTHDAY, PET
STEMVESTIGATION: DIGITAL SCALES

TECHNOLOGY

READ	Waxing Colorful
DISCUSS	Have you ever seen a parent or caregiver use a scale to weigh themselves or to weigh food?
EXPLAIN	A scale measures the pull of objects to the planet based on gravity. When an object is placed on the tray, a flexible (bendy) piece inside the tool bends. A little piece of foil is pushed to change the flow of electrical current. A signal goes to the screen showing the object's weight on Earth. Basically, pressing the top sends a signal. More pressing means more weight.
MOVE	Locate 5 objects to weigh on the scale. (This helps them begin to estimate the weight of objects less than 5 pounds.)

INVESTIGATE	
\multicolumn{2}{|l|}{Use the MindBugs Digital Scale activities to weigh items around the classroom. Upon introduction to the tool, explain its use including all the technical terminology above. Also, show students how the scale works. Set it for ounces or grams as desired before having students work with the device. When they write down answers, students should absolutely write down the decimal and parts of a whole. Introducing the vocabulary now will help understanding later. Educators use extensive vocabulary in Language Arts, and schools should begin the same habit for STEAM subjects.}	
ELA	Keep an invention journal for the week. Draw or write about tools and technology noticed during the week!

HAPPY BIRTHDAY, PET
STEMVESTIGATION: DIGITAL SCALES

Use a digital scale to weigh 6 objects.

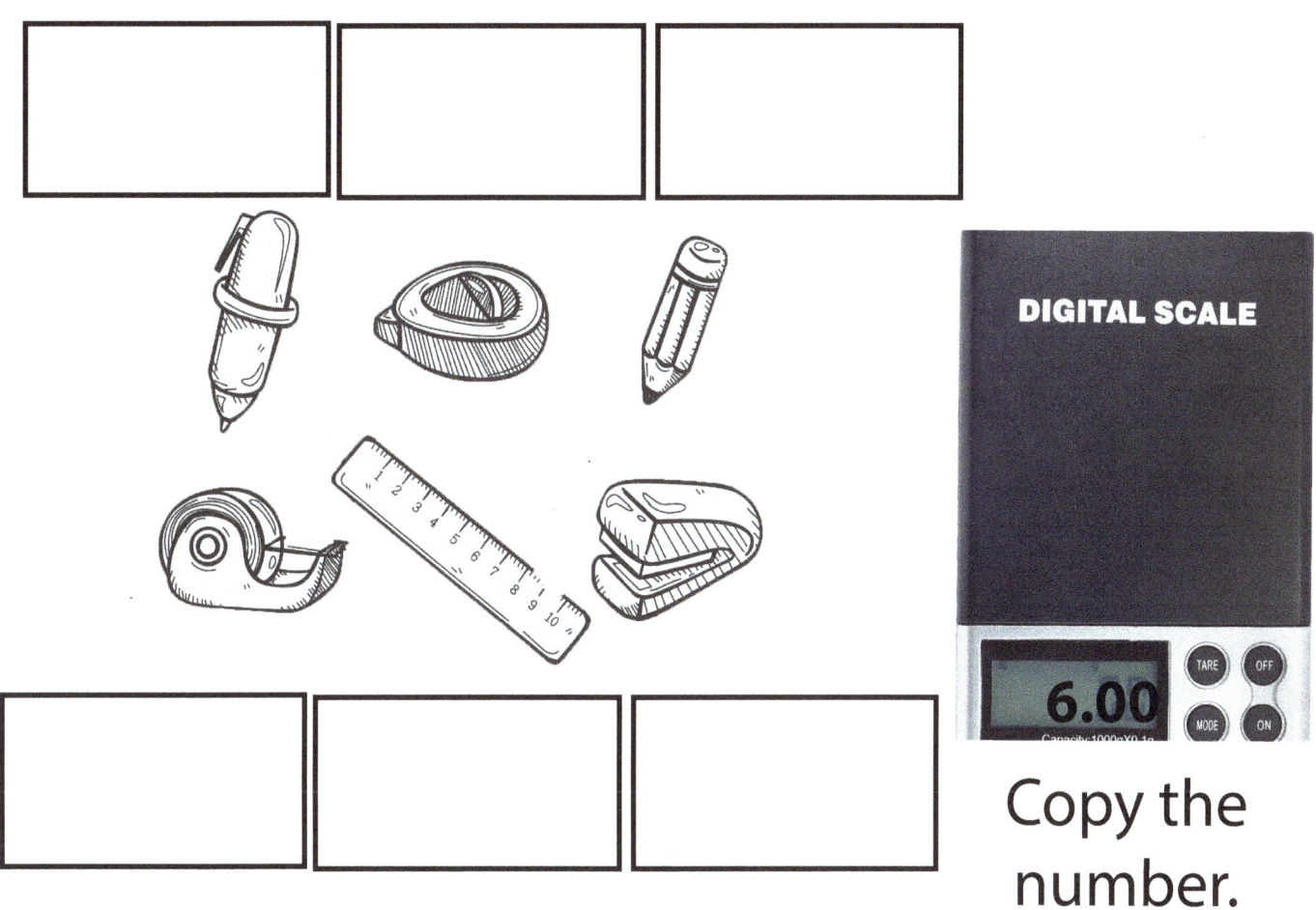

Copy the number.

Write the dot.

Waxing Colorful
Explore wax in technology.

If you wanted to build a house, what shapes would you use?

Waxing Colorful
Explore wax in technology.

In the hive (or in a wild nest), there are three types of bees:
1. a single female queen bee
2. up to 2000 male drone bees
3. some 20,000 to 40,000 female worker bees.

The worker bees raise larvae and collect the nectar that will become honey in the hive. When they leave the hive, they collect sugar-rich flower nectar and return.

Honey bees are a little like nurses, a little like architects, and a little like builders. They created wax to build honeycomb in regular hexagons.
The honey bees use a shape that fits together as well as a puzzle. There are no holes or gaps between hexagons in the honeycomb. No wax-making honey or energy is wasted. That also makes honey bees sort of like engineers learning the best way to do a job.

The honeycomb "house" idea is working for humans, too. A company in the United Kingdom builds hexagonal houses. Each room of the house fits tightly together to save resources. These houses work for people the way a honeycomb works for bees. The honeycomb shapes take less energy to build, heat and cool.

People have many uses for the beeswax that bees will not use. They make a lot more than any hive needs. Bee farmers harvest the rest for things like candles. Wax from honeybees is also used to make crayons. Most crayons are made of paraffin, but some organic companies are making crayons that contain fewer chemicals. Beeswax is mixed with pigments to make a rainbow of colors. You might be using beeswax candles or crayons in your house!

HAPPY BIRTHDAY, PET
STEMVESTIGATION: PERSPECTIVE DRAWINGS

ENGINEERING

READ	Strength in Numbers
DISCUSS	What will happen if I step on this tube? Elicit responses, then step evenly on the top of the tube. They may observe on their own that a slight lean in any direction causes the tube to crumple more quickly on that side. Placing a piece of cardboard over the top of the tube may help distribute weight more evenly.
EXPLAIN	Some structures are supported by columns. Patios, porches, basements, and bridges are places you may see columns. Sometimes you cannot see the triangles, because they are hidden inside the walls.
MOVE	Students try balancing items on their own tubes (paper towel or toilet paper) to test strength.

INVESTIGATE

Given construction paper or card stock, challenge teams to devise a column that -when stood vertically- holds 1 pound of weight. There are many possible solutions:

- Increase the stiffness of the sides of the tube to help the structure resist buckling under a load.
- Decrease the possibility of collapse by filling the tube. The load is distributed evenly by the material inside the tube. A column can be filled with cheaper material and still increase the column's compression strength.

Test the strength of columns with a 1-pound book. Chart tube thickness vs weight held. Chart material in the column vs weight held.

ELA	Explain the box pattern to a teammate.

HAPPY BIRTHDAY, PET
STEMVESTIGATION: PERSPECTIVE DRAWING

Draw the rest of the shape.

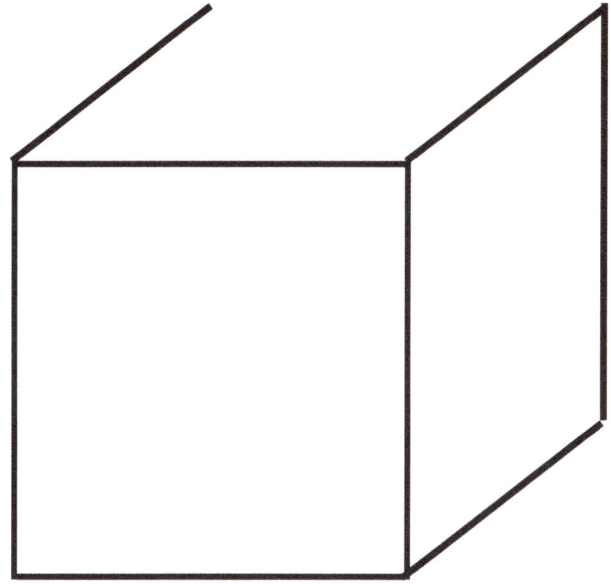

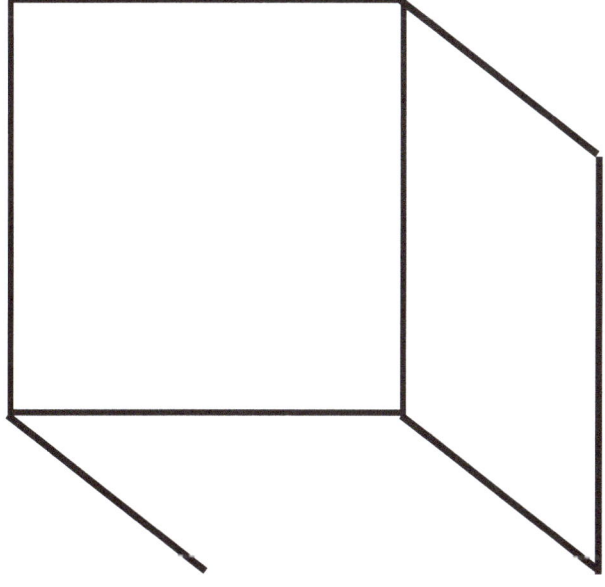

Strength in Numbers
Explore columns in engineering.

Visit http://alturl.com/aq5vy to make a Log Cabin.
Visit http://www.artistshelpingchildren.org for folded boats and lots of other great ideas.
Visit http://www.makingfriends.com to make hats in under a minute!
Visit http://www.creativekidsathome.com to make a dome.

Strength in Numbers
Explore columns in engineering.

Ginny kicked up dust with her sneakers as she headed down the dirt road. She was just wondering if the day could get any hotter when a motion caught her eye. Something was fluttering in the ditch to her left. She walked closer to get a look. She saw that it was most of a newspaper. Ginny did not want to leave litter in the road, so she rolled the paper into a tube and tucked it under her arm. She'd throw it away later.

She was about halfway to her friend's place when the sun broke free of the clouds. It was hot! It was sunny! The glare made her head ache. Ginny remembered the newspaper. She took a page and folded it into a paper hat. Her class had made them for a play last year. She never thought that skill would come in handy. Whew! Her face was cooler in the hat's shade.

Ginny pretended she was a pirate in her buccaneer's hat. She found a stick and heaved it around like a sword. Ginny was doing a pretty good pirate's "arrrr" when she crossed the little bridge over the Spring Road stream. Yesterday's rain had raised the water level, and there was a great current. Ginny stopped to unfold two pages of the newspaper. She folded two small paper boats and set them gently down in the water on the North side of the bridge. She ran to the South side of the bridge to see which boat would arrive first.

She turned back to the road and finished her walk to Meghan's house. Just as she turned into her best friend's driveway, Ginny saw Meghan's father looking for something in his car. Ginny was curious.

"What are you looking for, Mr. Harris?" asked Ginny curiously. He showed her a letter that needed an envelope. Ginny had a handy solution. She used the last page of her newspaper to fold a quick envelope for Mr. Harris. He uses a sticker to close the flap, added a stamp and mailed his letter.

He asked, "Why are you carrying that old newspaper around?"

Pirate Ginny answered, "Because, it's not just a newspaper..."

HAPPY BIRTHDAY, PET
STEMVESTIGATION: MOSAICS

READ	Square in the Middle
DISCUSS	Have you ever seen a mosaic? Has the art class ever created mosaics before? What do you think a mosaic might be?
EXPLAIN	According to WiseGeek.org, mosaic is an art form that involves fitting small pieces of rock, shell, tile, or glass called tesserae together to create a pattern that may be abstract or representational. In Rhombi's words, fit little shapes together to make a big shape.
MOVE	Create a people mosaic. Lie down to make a shape (flower or sun is easiest), and make sure to take a photo!

INVESTIGATE

Cut colored paper into one inch strips. Students will use age-appropriate scissors to cut strips down to small squares for use in the mosaic.

Students draw a very simple picture in white crayon on black construction paper. Then, working from the center to the outside, cover the image with colored squares. For each "tile" used, touch the tile to a glue-saturated sponge to minimize mess. Color and placement are all child's choice. Let them explore.

ELA	Use your mosaic to tell a picture story.

THE ARTS

HAPPY BIRTHDAY, PET
STEMVESTIGATION: MOSAICS

Make art with polygons. Use squares and triangles. Draw one of these shapes in each square.

Square in the Middle
Explore mosaic patterns in art.

Square in the Middle
Explore mosaic patterns in art.

Quad sat glumly on the bench. There was a box on the ground between his feet. When Rhombi sat down, she looked in the box.

"Quad, why are you hauling a box of broken pottery and dishes around the school?"

Quad looked into the box and put his head in his hands. His elbows rested on his knees, and he groaned. "That was my birthday present for Grandad. I dropped it on the way out of the art room, and I don't have enough money to buy another one."

Rhombi pulled the box closer and pushed some of the shards with her pencil. She closed her eyes and began to brainstorm. What could you do with a box and broken plates? Rhombi thought about broken things. Where had she seen broken plates in the last week?

"I have it," cried Rhombi loudly. She giggled and said, "Sorry, that was really loud, but I have an idea." Quad's friend told him about the table her parents had bought for the patio. It had a pattern glued to the top in the shape of the moon and stars. The artist had glued each piece down and filled the spaces with ??.

"You could do the same thing with this box," said Rhombi. "Draw a design, and fill it in with the broken pieces."

Quad sat up and grinned. He liked the idea. Grandad loved cats. Quad could draw a cat on the box. He grabbed the box and headed toward the art room. Turning, he said to Rhombi, "You are a genius!" She smiled and waved him away.

Quad drew a cat. He glued all the broken pieces to the box. The art teacher gave Quad some grout to fill in the cracks between pieces. Mr. Kandinsky said that the grout would be dry in 24 hours. Quad left his masterpiece in the art room window. He was happy with the gift he'd give Grandad, and he knew Grandad would love that he'd made the present all by himself.

HAPPY BIRTHDAY, PET
STEMVESTIGATION: LENGTH

READ	Comin' Up a Cloud
DISCUSS	Are you taller or shorter than this stick? What might be wider than this stick? How far can you count on the stick? What is the highest number you recognize on the stick?
EXPLAIN	This stick is a yardstick that has 3 big sections and 36 little sections. The 3 are called feet. The 36 are called inches. Today we'll use the yardstick to measure our string. Each piece of string will be one yard, or 3 feet or 36 inches long.
MOVE	Have everyone mark their heights on the whiteboard. and compare to the yardstick. Line up for a conga line in ascending height order. Count to 36 (# inches on yardstick).

INVESTIGATE

Place string along the length of the yardstick. Cut string to the same length as the stick. Using the string and glue, create a shape picture. You'll know that your finished shape measures exactly one yard, 3 feet or 36 inches!

ELA	Students explain the process of creating their present their shapes to older students.

MATHEMATICS

HAPPY BIRTHDAY, PET
STEMVESTIGATION: LENGTH

Measure string to make line art.

Lay string along the length of a measuring stick. Cut 36 inches (3 feet) of string. Use your string to make a cool shape. Glue the shape to paper.

Comin' Up a Cloud
Explore materials in mathematics.

A cotton boil weighs about 4 grams. A paper clip weighs about 1 gram, so a cotton boll is equal in weight to about 4 paper clips.

Comin' Up a Cloud
Explore materials in mathematics.

Ginny carefully wove through the rows of cotton. The Georgia weather was hot and humid. Ginny felt like her clothes were sticky, and the tiny gnats made mad little dashes toward her face and ears. She fanned them away. They didn't bite, but the tiny flies were annoying.

She'd kept an eye on the cotton since planting season in February. The flower buds had opened, and she'd watched as the petals fell to the dusty ground. With petals gone, the rest of the plant had ripened and grown. Each plant had a lot of cotton "bolls" with fluffy fibers.

This week, she and her family would hire local help to harvest the bolls. Then the piles of not-so-fluffy cotton would be "ginned" to get rid of all the seeds. They would send the cotton off to a mill. The mills will mix and clean the cotton. Picking machines will break the piles of cotton into pieces. Dirt has to be shaken out of the cotton fibers.

A carding machine that separates the fibers will comb through the piles. Combing makes sure all the fibers run in the same direction. Ginny's Great-Great Grandma used to do all this by hand, but the machines work with a lot more fiber in a much shorter amount of time. Great-Great Grandma spent weeks doing what a machine finishes in an hour. A long, smooth rope called a sliver forms. The sliver has to be pulled and twisted until it's thin. The thin "roving" is finally ready for a spinning frame. Before it becomes string, all that beautiful roving is twisted and wound onto bobbins.

Ginny wondered where the cotton boil she touched would go. Would it end up on a cargo ship bound for China? Would it land in a pretty gift store? Would she wear it next year as a t-shirt or a pair of jeans? Maybe it would wind her yo-yo or fly a kite. As Ginny bent down to tie her cotton shoelace, a cloud crossed the field. She glanced at the sky. It looked like it was coming up a cloud.

Read more : http://www.ehow.com/about_6360791_cotton-made-thread_.html STEAMStart 2D / 3D SHAPES KINDERGARTEN ALL RIGHTS RESERVED ©2014 1O80 EDUCATION, INC.

HAPPY BIRTHDAY, PET
PLAY WITH SHAPES

HOPSCOTCH

MATERIALS
chalk
surface safe for drawing and jumping
cut sponges (or rocks if you are feeling particularly bold)

SETUP
Draw hopscotch boards for young children.
Encourage older students to draw their own game boards.

INSTRUCT THE PLAYERS
The first player tosses a marker (rock, coin) into the first square; it must land within the confines of the square without bouncing out or touching a line. The player then hops through the course, making sure to skip the square with the marker in it. Players hop in single squares with one foot (either foot is fine), and use two feet for the side by side squares, one in each square. Upon completion of the hop sequence, the player continues her turn, tossing the marker into square number two and repeating the pattern. Players begin their next turn where they last left off. Player loses a turn (or ends the current turn) if these things happen:
- player steps on a line;
- player misses a square on toss;
- or the player loses balance.

INSTRUCT THE PLAYERS
Explain the rules.
Demonstrate one full turn.
Have students announce each number as it is hopped, counting up to 10 and back down to 1.

Rhombi's Adventures in 3D

HAPPY BIRTHDAY, PET
UNIT 9 CHALLENGE

HAPPY BIRTHDAY, PET CHALLENGE

Once upon a time there was Rhombi. Rhombi lived in a house shaped like a cube with a cool pyramid roof. The house would not be toppled by Autumn winds. The pointed pyramid roof kept out Winter rain and snow.

HAPPY BIRTHDAY, PET CHALLENGE

Rhombi's pet had a room of its own with 8 corners (vertices) where Pet stored her treasures. Pet's place stayed dry in the Spring rain and cool in the Summer sun.

Where is Pet?

HAPPY BIRTHDAY, PET CHALLENGE

Rhombi's was planning Pet's birthday party and wanted to invite many friends. Rhombi and Pet drew party invitations on circus paper. Pet wanted 8 animal friends to attend her party.

HAPPY BIRTHDAY, PET CHALLENGE

Rhombi realized that she didn't have enough room for a house full of friends. She decided to ask for help. She looked beyond the pages of her world and asked, "Designers, can you help me make a space for Pet's friends?"

HAPPY BIRTHDAY, PET
TALK ABOUT STEAM

What 3D shapes are part of Rhombi's house? **cube, pyramid**

How many edges does a pyramid have? **8 edges**

Rhombi lived through what two seasons in her new house? **Fall and Winter**

Plan a party! What will you eat? How will you decorate?

If you have a party budget of $20, how will you spend the money?
Consider using catalogs or ads to actually plan the menu and budget.

What types of animal friends might live in the forest?
birds, rabbits, foxes, turtles (any appropriate animals or insects)...

Will you design a party porch? Will you design a whole new room?

What materials will you need in order to build Rhombi's pet's house?
Answers will vary.

RHOMBI'S ADVENTURES IN 3D
UNIT 4 CHALLENGE: HAPPY BIRTHDAY, PET

"Designers, will you help me create a great party space for Pet?"

my idea	my plan

On the back of this paper, draw the steps you will need to take.

RHOMBI'S ADVENTURES IN 3D
UNIT 4 RUBRIC: HAPPY BIRTHDAY, PET

TASK Students will use their knowledge of habitat, house, pyramids and strength to build a scale model of a celebration space for pets.

	Content	Organization	Design & Build
1	• Is well thought out and supports the solution to the challenge or question • Reflects application of critical thinking • Has clear goal that is related to the topic • Is accurate	• Information is clearly focused in an organized and thoughtful manner • Information is constructed in a logical pattern to support the solution	• Uses 2D shapes to create 3D shapes • Neatly joins corners and edges • Constructs a platform with strong pillars able to withstand given weight
2	• Supports the solution • Has application of critical thinking that is apparent • Has no clear goal • Has some factual errors or inconsistencies (i.e. wheels do not rotate around axle)	• Project has a focus but might stray from it at times (more concerned with form than function) • Information appears to have a pattern, but the pattern is not consistently carried out in the project	• Edge seams are mis-joined and uneven • Constructs an adequate platform that does not support weight
3	• Provides inconsistent information for solution • Has no apparent application of critical thinking • Has no clear goal • Has significant factual errors, misconceptions, or misinterpretations	• Content is unfocused and haphazard • Information does not support the solution to the challenge or question • Information has no apparent pattern	• Edge seams are missing or ragged • Student project falls apart and student is unable to make changes

Rhombi's Adventures in 3D

RHOMBI'S CANDIES
EXTENSION

CURRICULUM CONNECTIONS

Rhombi is opening a candy cart, and she needs help with the design.

ART CONNECTIONS

A logo is a piece of art that uses graphics to convey a message or help consumers recognize a product. The logo is a design that includes a name & symbol or acronym identifying and/or distinguishing your brand from others. A logo uses a SYMBOL (emblem, icon, sign) in its DESIGN (the "look and feel" including lettering, color, shape, idea, pattern).

TECHNOLOGY CONNECTIONS

Clipart Collections
Http://school.discoveryeducation.com/clip-art/

HISTORICAL CONNECTIONS

Some of the world's current businesses have stood the test of time. Lloyd's of London started life in 1688 as Edward Lloyd's Coffee House. Japan's Kongo Gumi temple builders had a 1,428 year run that ended in 2006.

The first official logo was trademarked in 1876 by Bass Ale, but individuals and companies have used personal "seals" for thousands of years. A carved stamp was pressed into hot wax so recipients would know that the origin of the document was authentic.

MATH BACKGROUND

As explained by www.k-5mathteachingresources.com, the number line was originally proposed as a model for addition and subtraction by researchers from the Netherlands in the 1980s. A number line is a visual representation for recording and sharing students' thinking strategies during the process of mental computation. Students will begin to solve problems mentally by picturing the number line in their heads. A number line is a line in which real numbers can be placed, according to their value. Each point on a number line corresponds to a real number. Each real number has a unique point that corresponds to it. Ex: the number 2.5 (2 1/2) corresponds with the point on a number line that is halfway between two and three.

MINDBUGS TO NOTE

- Students confuse the minute and hour hands.
- Students have difficulty estimating the duration of a given length of time.
- Digital clocks and timers have number scales based on 60 not 100.
- Students often forget to begin measuring a length from the zero mark.
- They use the edge of the ruler or start at 1.
- When measuring lengths longer than the ruler, some students flip the ruler over and over.

CURRICULUM CONNECTIONS

COMMON CORE CONNECTIONS

ELA/Literacy
RI.K.1 With prompting and support, ask and answer questions about key details in a text. W.K.1 Use a combination of drawing, dictating, and writing to compose opinion pieces in which they tell a reader the topic or the name of the book they are writing about and state an opinion or preference about the topic or book. W.K.2 Use a combination of drawing, dictating, and writing to compose informative/explanatory texts in which they name what they are writing about and supply some information about the topic.

Mathematics
MP.2 Reason abstractly and quantitatively. MP.4 Model with mathematics. K.CC.A Know number names and the count sequence. K.MD.A.1 Describe measurable attributes of objects, such as length or weight. Describe several measurable attributes of a single object. MD.A.1 Describe measurable attributes of objects, such as length or weight. Describe several measurable attributes of a single object.

NEXT GEN SCI CONNECTIONS

ETS1.A: Defining Engineering Problems
A situation that people want to change or create can be approached as a problem to be solved through engineering. Such problems may have many acceptable solutions.

1-ESS1-1,1-ESS1-2 Patterns in the natural world can be observed, used to describe phenomena, and used as evidence.

ESS2.D: Weather and Climate
Weather is the combination of sunlight, wind, snow or rain, and temperature in a particular region at a particular time. People measure these conditions to describe and record the weather and to notice patterns over time. (K-ESS2-1)

K-ESS2-1. Use and share observations of local weather conditions to describe patterns over time.

UNIT S.T.E.A.M. ACTIVITIES

Rhombi Audio Download / Video (available December 2014)

Extension:	Rhombi's Candies:	Science:	Melting
Extension:	Rhombi's Candies:	Technology:	Tools
Extension:	Rhombi's Candies:	Engineering:	Reverse Engineering
Extension:	Rhombi's Candies:	Art & Music:	Graphic Design
Extension:	Rhombi's Candies:	Mathematics:	Charts and Calendars
Extension:	Rhombi's Candies:	ELA:	Read and illustrate "Rhombi's Candies."

BUILD S.T.E.A.M. WITH GREAT BOOKS

SCIENCE

The Candymakers
by Wendy Mass

Classroom Library
candy
chocolate
melting
cooking

TECH

Learn about the origins of chocolate!
http://www.exploratorium.edu/chocolate/live.html

Classroom Library
time
telling time

ENGINEERING

How Willy Got His Wheels
by Deborah Turner

The Box with Red Wheels
by Maud Petersham

Classroom Library
carts
wagons
wheels

THE ARTS

The Wheels on the Bus
(Raffi Songs to Read)
by Raffi

Classroom Library
starting a business
making candy

MATH

Classroom Library
charts and graphs
stopwatches
time

RHOMBI'S CANDIES
STEMVESTIGATION: MELTING

READ	**HOW CANDY IS MADE** HTTP://www.hersheys.com/ads-and-videos/how-we-make-chocolate.aspx
DISCUSS	What are the steps for making chocolate? What is melting? What is your favorite candy?
EXPLAIN	Chocolate makes a physical change at 86 degrees Fahrenheit. That is cooler than your body temperature. The chocolate melts as it turns from a solid to a liquid. Sunlight is usually hot enough to melt chocolate. Turn chocolate back into a solid by putting it into a refrigerator or freezer.
MOVE	Act out melting and freezing. Drop into a puddle on the floor as one child plays sunlight. Another child swoops in with cold weather, and kids stiffen.

INVESTIGATE
Challenge kids to melt chocolate in unusual ways. Will a lamp melt the chocolate enough to mold the chocolate in a candy mold? Will sunlight be enough to melt the chocolate?

ELA	Each hour of the day, take digital photos of processes.

SCIENCE

HAPPY BIRTHDAY, PET
STEMVESTIGATION: TIME TOOLS

READ	Mindbugs in Time: Elapsed Time *Appendix*
DISCUSS	How many students have participated in sports? Did you see a referee using a stopwatch? What about spelling bees or other activities that must be timed.
EXPLAIN	Stopwatches measure elapsed time, the number of seconds that pass from start to finish of an activity. (Demonstrate the stopwatch.)
MOVE	How many jumping jacks can you do in a minute? How many seconds pass while you do 10 lunges?

INVESTIGATE

Use the stopwatch while lining up for breaks, during bathroom trips, lunchroom cleanup, etc. Choose a set of activities from the Mindbugs Stopwatch Activity Guide. Challenge students to master the stopwatch while timing everything from jumping jacks to repeating the alphabet.

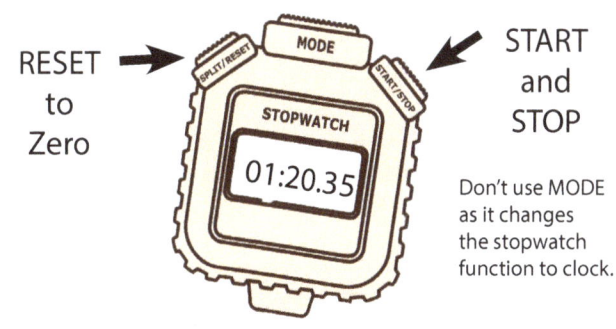

RESET to Zero

START and STOP

Don't use MODE as it changes the stopwatch function to clock.

ELA	Demonstrate the stopwatch to another student!

TECHNOLOGY

HAPPY BIRTHDAY, PET
STEMVESTIGATION: REVERSE ENGINEERING

EXPLORE	Create a Car Interactive Game http://www.abcya.com/create_and_build_car.htm
DISCUSS	What parts are needed to make a vehicle? What troubles did you have (your team have or the whole group have)?
EXPLAIN	Vehicles that roll must have an axle. It is the rod that holds wheels and lets them roll. Wheel and axle together are machines we use to make life easier.
MOVE	Play people carts. One student lays on stomach. Second student takes a firm grip on the prone student's ankles. Lifting the student by the ankles makes a human cart.

INVESTIGATE	
Set up a reverse engineering table in the classroom. Students take apart all kinds of wheeled vehicles using hand tools. Toy cars are inexpensive at the local Goodwill, or ask for donations using one of the local news options.	
ELA	Draw the process of disassembling, and include the parts.

ENGINEERING

RHOMBI'S CANDIES
STEMVESTIGATION: COLOR

READ	Mouse Paint by Ellen Stoll Walsh
DISCUSS	What colors did the mice discover? How did the mice paint? What is your favorite color?
EXPLAIN	Primary colors may be combined to create secondary colors. Red and blue make purple. Blue and yellow make green. Red and yellow make orange. Colors change because some light bounces back from an object and our eyes see the color differently.
MOVE	Call out a secondary color. Kids with the two primary colors that are found in the second color find each other. Call out another color, etc.

INVESTIGATE
Using water and licorice, create art. Place licorice of various colors on the tray with a shallow bowl of water. Leave cotton balls, paint brushes, sponges, and other items that kids might use in painting.
Dragging colored licorice across white paper produces interesting "string" art. Challenge kids to match some color palette created by you. What amount of water and colors of licorice recreate your palette?

ELA	Draw a "recipe" for the ratio of candy to water used to recreate the color palette.

THE ARTS

RHOMBI'S CANDIES
STEMVESTIGATION: CHARTS AND GRAPHS

READ	Anna Banana (jump roping rhymes) Joanna Cole
DISCUSS	Charts and graphs are things we've used this year to collect information. What projects used charts or graphs this year? What information could you collect using a chart?
EXPLAIN	Charts and graphs are math problem-solving tools. They all follow a sort of pattern so that everyone may read the story they tell.
MOVE	Jump rope, and count the number of full turns.

INVESTIGATE
Make a chart of jump rope numbers. Use shirt color to group kids on the chart. Talk about reasons that one color or another may have a taller bar on the graph. Does shirt color cause better jump roping? If you wear a t-shirt of another color, will your ability to jump grow? Try it. Chart it. Talk about other reasons (variables) that might be causing differences in height on the chart.

ELA	Take part in the discussion, and offer personal observations regarding variables.

RHOMBI'S PLACE
PLAY WITH SHAPES

PIC SHAPES

MATERIALS
whiteboard and whiteboard markers
or
markers and large paper

SETUP
Use geometry words and vocabulary from all subjects or attribute cards you already own to fill a bag with options.

INSTRUCT THE PLAYERS
Break the class into two teams. Team one draws a word to be illustrated. One player draws while others guess the word.

Team two must keep the same word and find a different way to represent its meaning. THEN, pull a new word to illustrate.

Play ends after 20 minutes.

Rhombi's Adventures in 3D
RHOMBI'S CANDIES
EXTENSION CHALLENGE

RHOMBI'S CANDIES CHALLENGE

Once upon a time, there was Rhombi. Rhombi had a wonderful friend named Pet and a cool cube playhouse with a strong pyramid roof. Pet even had her own house and a party pavilion for friends!

RHOMBI'S CANDIES CHALLENGE

Pet and Rhombi loved 2 dimensional and 3 dimensional shapes of all kinds. Their house was filled with interesting squares, triangles, hexagons, cubes, and pyramids. Rhombi's favorite project was a circle tree hanging on her wall.

RHOMBI'S CANDIES CHALLENGE

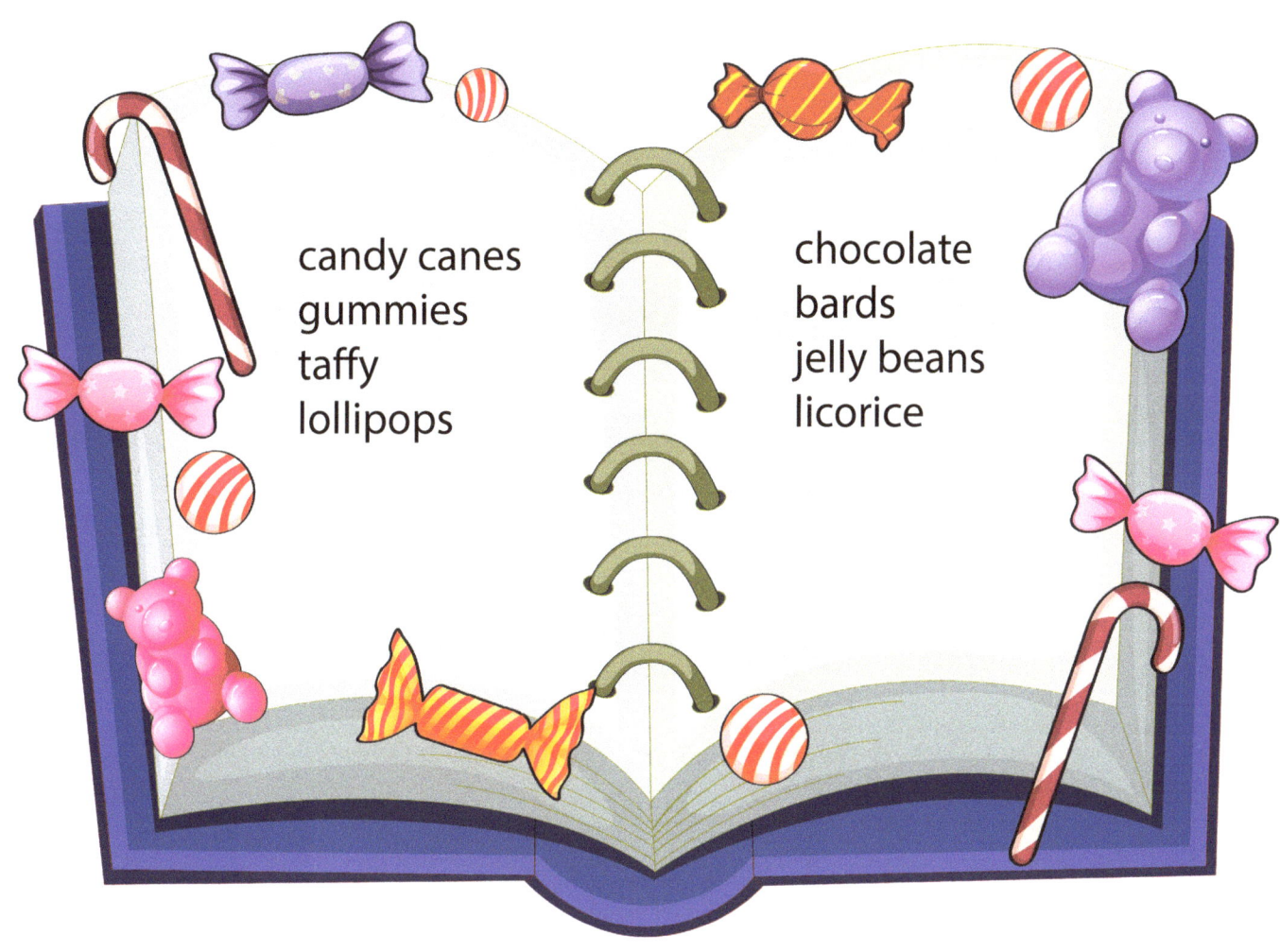

candy canes
gummies
taffy
lollipops

chocolate
bards
jelly beans
licorice

Rhombi's friend Pet had celebrated her birthday just a few months ago. They had served circus food including cotton candy, lollipops, popcorn, and other pet friendly candy. It had been so popular that Rhombi wanted to recreate the menu at the market.

RHOMBI'S CANDIES CHALLENGE

Rhombi's parents had given permission for Rhombi to open a candy store cart. The cart could be wheeled down to the Saturday farmer's market. When the market ended, the cart could be wheeled right back home.

RHOMBI'S CANDIES CHALLENGE

The farmer's market hosted lots of different foods and crafts. Rhombi's friends Pi and Radius sold fruits and vegetables at the market. She had asked to borrow the cart one weekend and try her business idea.

RHOMBI'S CANDIES CHALLENGE

Pi and Radius had to visit Grandma, so they loaned Rhombi their cart one weekend. She had piled it high with candy and cookies.

RHOMBI'S CANDIES CHALLENGE

The cart was open, and all the chocolate melted. Rhombi decided that she needed a new design for her candy cart. Rhombi looked beyond the pages of her world and asked, "Designers, will you help me create a candy store cart with the materials we have?"

RHOMBI'S CANDIES
EXTENSION CHALLENGE: MAKE A PLAN

Draw at least 3 ways Rhombi might move her candies to the market.

RHOMBI'S ADVENTURES IN 3D
EXTENSION CHALLENGE: RHOMBI'S CANDIES

"Designers, will you create a candy cart for the Saturday Market? The cart should move 2 feet, and it should keep 3 pieces of chocolate from melting under a heat lamp."

my idea

my plan

RHOMBI'S CANDIES
TALK ABOUT STEAM

What businesses does a town need to support townspeople?
ex: grocery story or farmer's market, town hall, police department, school, businesses

What would you add to a town to make it really special? ex: ice cream shop, etc.

Make a list of 2D shapeswith 3,4,5,6,7,8,9,and 10 sides.
triangle, square (quadrilateral), pentagon, hexagon, heptagon, octagon, nonagon, decagon

Make a list of 3D shapes.
cube, pyramid, prism, cylinder, sphere, etc.

What made Rhombi want to open a candy shop? Pet's party candy

Where will Rhombi sell her candies? market

How will she move candy to and from the market? She will need a cart of some kind.

What steps will you take in making Rhombi's cart? Answers will vary.

RHOMBI'S ADVENTURES IN 3D
EXTENSION RUBRIC: RHOMBI'S CANDIES

TASK: Students will use their knowledge of motion, shapes, and melting to create a canopied market cart.

	Content	Organization	Design & Build
1	• Is well thought out and supports the solution to the challenge or question • Reflects application of critical thinking • comes up with a plan to move the cart from place to place whether or not the idea works	• Information is clearly focused in an organized and thoughtful manner • Information is constructed in a logical pattern to support the solution	• Uses 2D shapes to create 3D shapes • Neatly joins corners and edges of the cart. • Has a cart that shows evidence of a plan for motion
2	• Supports the solution • Has application of critical thinking that is apparent • Has no clear goal	• Project has a focus but might stray from it at times (more concerned with form than function) • Information appears to have a pattern, but the pattern is not consistently carried out in the project	• Edge seams are mis-joined and uneven • Constructs an adequate platform that will not stay intact if filled with candy
3	• Provides inconsistent information for solution • Has no apparent application of critical thinking • Has no clear goal • Has significant factual errors, misconceptions, or misinterpretations	• Content is unfocused and haphazard • Information does not support the solution to the challenge or question • Information has no apparent pattern	• Edge seams are missing or ragged • Student project falls apart and student is unable to make changes

APPENDIX

POLYGON PATTERNS AND VOCABULARY

POLYGON NET PATTERNS	193-194
POLYGON TRACING SHAPES	195-203

ASSESSMENTS

PRE AND POST ASSESSMENTS	204-212
SCOPE AND SEQUENCE	213
PLANNING PAGES	214-215
VOCABULARY	2016-2017

CORRELATIONS, ASSESSMENTS, AND REFERENCE MATERIAL

COMMON CORE CORRELATIONS	218-220
NEXT GEN SCIENCE STANDARDS	175

APPENDIX
CUBE NET

INSTRUCTIONS: Print this net for use in a center or during STEMVESTIGATIONS.

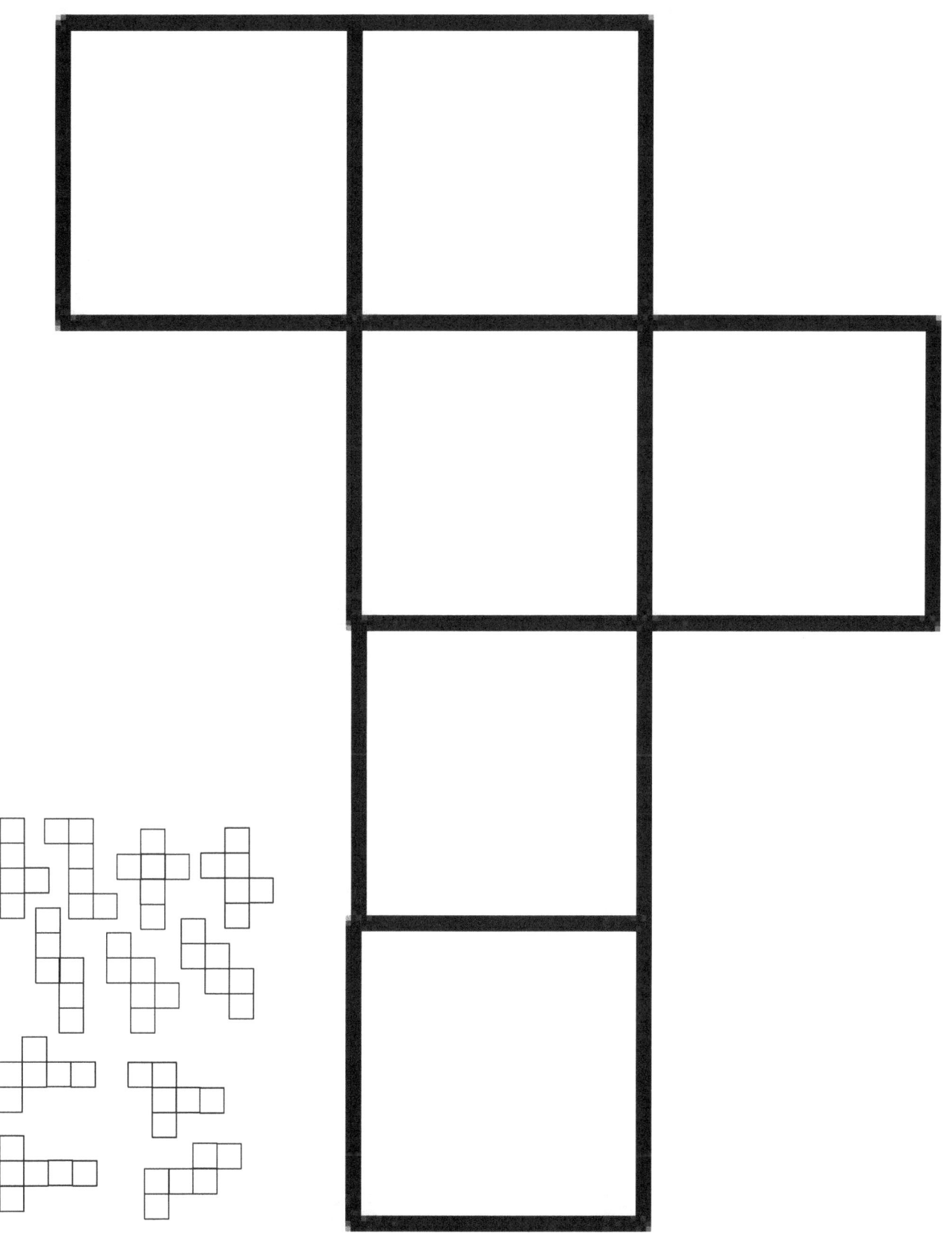

STEAMStart copyright 2020 by Jeannie Ruiz All Rights Reserved

APPENDIX
PYRAMID NET

INSTRUCTIONS: Print this net for use in a center or during STEMVESTIGATIONS.

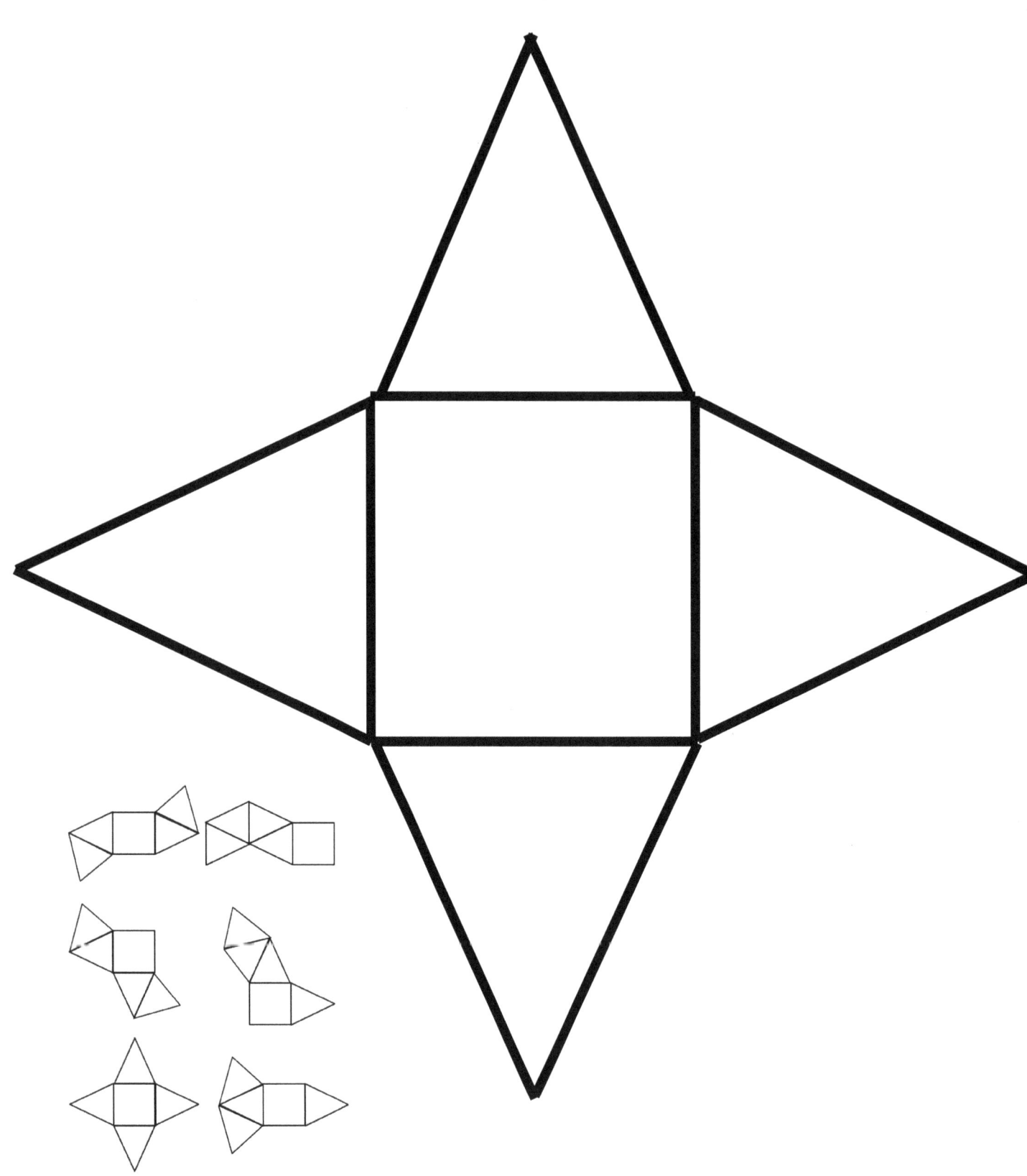

RHOMBI'S POLYGON ROUND-UP
CIRCLE

ABOUT THE CIRCLE
- In the math world, circle refers to the boundary of the shape. "Disk" is used to refer to the whole shape, including the inside.
- A straight line from the center of a circle to the edge is called the radius.
- A straight line that passes from one side of a circle to the other through the center is called the diameter.
- The distance around the outside of a circle is called the circumference. All points on the edge of a circle are the same distance to the center.
- The value of Pi (π) to 2 decimal places is 3.14, it comes in handy when working out the circumference and area of a circle.
- Circles have a high level of symmetry.

RHOMBI'S POLYGON ROUND-UP
TRIANGLE

ABOUT THE TRIANGLE
- Triangles are polygons with the least possible number of sides (three).
- The three internal angles of a triangle always add to 180 degrees.
- An equilateral triangle has three sides of equal length and three equal angles.
- The longest side of a right angle triangle is called the hypotenuse, it is always found opposite the right angle.
- Trigonometry is the study of the relationship between the angles of triangles and their sides.
- Triangle shapes are often used in construction because of their great strength.

RHOMBI'S POLYGON ROUND-UP
RECTANGLE

ABOUT THE SQUARE
- A square is a polygon with 4 sides of equal length and 4 right angle corners (90 degree corners).
- Because it has 4 sides of equal length, a square is a regular quadrilateral.
- A square is also a rectangle with equal sides and a rhombus with right angles.
- The perimeter of a square is 4 times the length of one side.
- Opposite sides of a square are parallel.
- The internal angles of a square add to 360 degrees.
- A square has 4 lines of reflectional symmetry.

RHOMBI'S POLYGON ROUND-UP
PENTAGON

ABOUT THE PENTAGON
- A pentagon is a 5 sided polygon with interior angles that add to 540 degrees.
- Regular pentagons have sides of equal length and interior angles of 108 degrees.
- The US Department of Defense headquarters is named 'the Pentagon'.
- The edible plant okra is shaped like a pentagon.

RHOMBI'S POLYGON ROUND-UP
HEXAGON

ABOUT THE HEXAGON
- A hexagon is a 6 sided polygon with interior angles that add to 720 degrees.
- Regular hexagons have sides of equal length and interior angles of 120 degrees.
- Beehive cells are hexagonal.

RHOMBI'S POLYGON ROUND-UP
HEPTAGON

ABOUT THE HEPTAGON
- A heptagon is a 7 sided polygon with interior angles that add to 900 degrees.
- Regular heptagons have sides of equal length and interior angles of 128.57 degrees.
- The British 50 and 20 pence coins are curved heptagons.

RHOMBI'S POLYGON ROUND-UP
OCTAGON

ABOUT THE OCTAGON
- An octagon is an 8 sided polygon with interior angles that add to 1080 degrees.
- Regular octagons have sides of equal length and interior angles of 135 degrees.

RHOMBI'S POLYGON ROUND-UP
NONAGON

ABOUT THE NONAGON
- A nonagon is a 9 sided polygon with interior angles that add to 1260 degrees.
- Regular nonagons have sides of equal length and interior angles of 140 degrees.

RHOMBI'S POLYGON ROUND-UP
DECAGON

ABOUT THE DECAGON
- A decagon is a 10 sided polygon with interior angles that add to 1440 degrees.
- Regular decagons have sides of equal length and interior angles of 144 degrees.

RHOMBI'S ADVENTURES IN 2D
PRE AND POST ASSESSMENT PAGES

PRINT FOR EACH STUDENT

RECOGNIZE SHAPES	179
REPRODUCE SHAPES	180
MOVE FROM 2D TO 3D	181
UNDERSTAND EQUALITY	182
MOTOR SKILLS	183

PRINT ONE SET FOR TEACHER

NUMERACY, CARDINALITY	184
OBJECT PERMANENCE	185
RUBRIC FOR NUMERACY, CARDINALITY, OBJECT PERMANENCE	186

RHOMBI'S ADVENTURES IN GEOMETRY
RECOGNIZE SHAPES

Color the square.	□	⬠	△	○
Color the triangle.	□	⬠	△	○
Color the rectangle.	▭	⬠	△	○
Color the circle.	□	⬠	△	○
Color the pentagon.	▭	⬠	△	○

RHOMBI'S ADVENTURES IN GEOMETRY
REPRODUCE SHAPES

Draw a square.	
Draw a triangle.	
Draw a rectangle.	
Draw a circle.	
Draw a pentagon.	

RHOMBI'S ADVENTURES IN GEOMETRY
MOVE FROM 2D TO 3D

Mark the shape you see on a cube.

 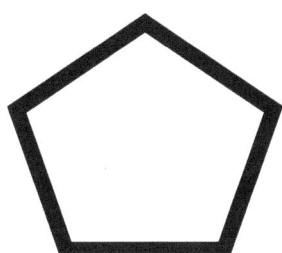  

Mark the shape you see on a pyramid.

Mark the shape you see on a the end of a cylinder.

 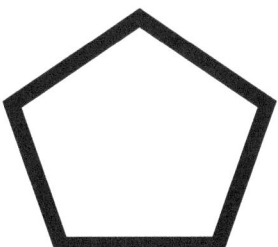  

RHOMBI'S ADVENTURES IN GEOMETRY
UNDERSTAND EQUALITY

Draw a line to break each shape into two equal parts.

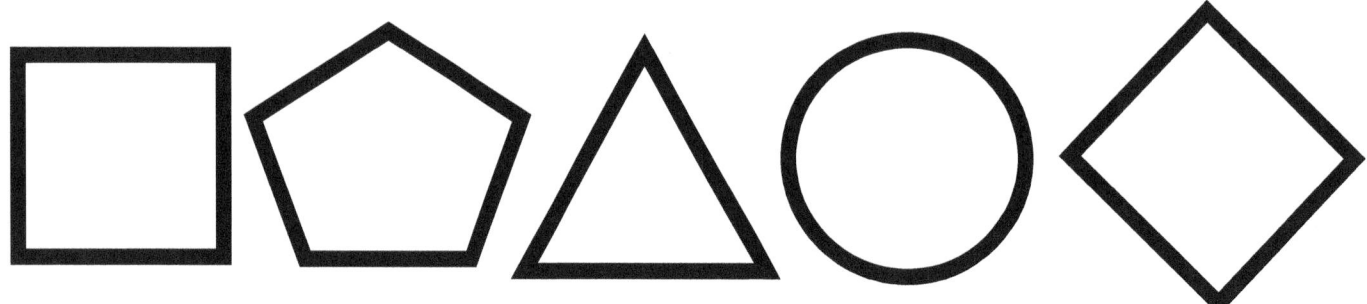

Draw the other half of each shape.

RHOMBI'S ADVENTURES IN GEOMETRY
RECOGNIZE SHAPES IN THE WORLD

Glue the cut shapes to this page to make a picture.

RHOMBI'S ADVENTURES IN GEOMETRY
ASSESS NUMERACY & CARDINALITY

DO NOT COPY FOR STUDENT
one-on-one with teacher

Show the child 2 pennies. Ask, "How many objects are in my hand?"

Child immediately says, "2." Has solid sense of abstract number 2...
Ready for discussions without visuals...

Child counts out loud and answers "2." Has basic understanding of abstract number 2...
Ready for handouts...

Child points to each object, counting out loud and answers "2."
Needs concrete objects to count while transitioning to abstract number sense...

Repeat actions for numbers 3 and 4.

Recognizing the sum of a pile of objects beyond 4 is almost impossible. Try it yourself. You will reflexively break the larger number into two or more lesser groups before combining to say the greater number.

All students will break numbers beyond 4 into smaller "piles." Note which students count by ones beyond 4 and which students automatically do some mental addition to give you the whole number.

RHOMBI'S ADVENTURES IN GEOMETRY
ASSESS OBJECT PERMANENCE

DO NOT COPY FOR STUDENT
one-on-one with teacher

Show the child your hand as you fold 2 pennies into your closed fist. Reopen the same hand. Ask. How many objects are in my hand?"

Child immediately says, "2." Has solid sense of object permanence for 2...
Ready for discussions without visuals...

Child counts out loud and answers "2." Has basic understanding of object permanence for 2. Ready for handouts...

Child points to each object, counting out loud and answers "2."
Needs concrete objects to count while transitioning to object permanence...

Repeat actions for numbers 3 - 9.

Remember, in this assessment, your goal is to determine a student's grasp of object permanence - NOT ability to count. The outcome is dependent on the student's ability to restate the cardinal number.

RHOMBI'S ADVENTURES IN GEOMETRY
ASSESS NUMERACY & CARDINALITY

Print rubric for your records. Perform assessment as early as possible.

STUDENT NAME	1	2	3	4	5	6	7	8	9

NOTE: All students will need to break amounts into smaller groups in order to recognize the cardinal numbers beyond "4." The brain naturally grasps groups of 1,2,3,4. It's why phone numbers are broken into smaller groups.

RHOMBI'S ADVENTURES IN GEOMETRY SCOPE AND SEQUENCE

	WEEK 1	WEEK 2	WEEK 3	WEEK 4	WEEK 5	WEEK 6	WEEK 7	WEEK 8
NEW HOME	**********	*********						
WEATHER	ONGOING							
DESIGN	45 MIN	CHALLENGE BRAINSTORM DESIGN BUILD TEST & USE DATA TO MAKE DESIGN CHANGES						
STRUCTURES	60 MIN							
CUBISM	45 MIN							
PERSPECTIVE	30 MIN							
ON THE ROOFTOP			*********	*********				
WATER AND SUN			45 MIN	CHALLENGE BRAINSTORM DESIGN BUILD TEST & USE DATA TO MAKE DESIGN CHANGES				
PYRAMID			90 MIN					
ROOF HEIGHT			45 MIN					
MATERIALS			60 MIN					
EROSION			30 MIN					
NEW FRIEND					*********	*********		
HABITATS					90 MIN	CHALLENGE BRAINSTORM DESIGN BUILD TEST & USE DATA TO MAKE DESIGN CHANGES		
ATTRIBUTES					90 MIN			
FOUNDATIONS					45 MIN			
PATTERNS					45 MIN			
SIZE AND SCALE					45 MIN			
PET'S CELEBRATION							*********	*********
DESSERT							45 MIN	CHALLENGE BRAINSTORM DESIGN BUILD TEST & USE DATA TO MAKE DESIGN CHANGES
DIGITAL SCALES							45 MIN	
STRENGTH							60 MIN	
GRAPHICS							90 MIN	
TIMELINES							30 MIN	

RHOMBI'S ADVENTURES IN GEOMETRY
TEACHER'S PRINTABLE UNIT PLANNING PAGE

3D SHAPES: UNIT					
STEMVESTIGATION	DATE	TIME	WHAT I'VE GOT	WHAT I NEED	PLANNING NOTES
SCIENCE					
TECHNOLOGY					
ENGINEERING					
ARTS					
MATH					
CHALLENGE					

RHOMBI'S ADVENTURES IN GEOMETRY
TEACHER'S PRINTABLE CHALLENGE PLANNING PAGE

3D SHAPES: UNIT					
STEMVESTIGATION	DATE	TIME	WHAT I'VE GOT	WHAT I NEED	PLANNING NOTES
CHALLENGE:					
BRAINSTORM & HYPOTHESIZE					
DESIGN					
BUILD					
TEST					
ANALYZE DATA					

VOCABULARY

EARTH AND SPACE	
DATA	SCIENCE INFORMATION
CONCLUSION	THE LAST PART
PREDICT	TO SAY WHAT YOU THINK WILL HAPPEN
DESCRIBE	SAYING OR DRAWING WHAT YOU SEE, HEAR, TOUCH, TASTE OR SMELL
OBSERVE	TO SEE, HEAR, TOUCH, TASTE OR SMELL
INVESTIGATE	TO GATHER INFORMATION
RECYCLE	TO USE AGAIN, ESPECIALLY TO REPROCESS
DISPOSE	TO PLACE OR SET IN A PARTICULAR ORDER; ARRANGE
REUSE	TO USE AGAIN, ESPECIALLY AFTER SALVAGING OR SPECIAL TREATMENT OR PROCESSING.
GOGGLES	A PAIR OF TIGHT-FITTING EYEGLASSES, OFTEN TINTED OR HAVING SIDE SHIELDS, WORN TO PROTECT THE EYES FROM HAZARDS SUCH AS WIND, GLARE, WATER, OR FLYING DEBRIS
DESCRIPTIONS	
FIVE SENSES	TASTE, SMELL, SEE, HEAR, TOUCH
HAND LENSES	A HAND HELD MAGNIFYING GLASS
MASS	AMOUNT OF MATTER IN AN OBJECT OR SUBSTANCE, MEASURED IN GRAMS
MATTER	ANYTHING THAT HAS MASS AND TAKES UP SPACE
NOTEBOOK	A BOOK OF BLANK PAGES FOR NOTES
OBJECT	THING
PATTERN	SHAPES, COLORS, OR LINES PUT TOGETHER IN A CERTAIN WAY
PHYSICAL PROPERTIES	ATTRIBUTES THAT CAN BE OBSERVED AND MEASURED
RELATIVE	COMPARING TO
TEXTURE	
ENERGY	THE ABILITY TO DO WORK
FIVE SENSES	TASTE, SMELL, SIGHT, HEAR, TOUCH
LIGHT ENERGY	ENERGY THAT YOU CAN SEE
NIGHT AND DAY	
PATTERN	EVENTS THAT REPEAT
DAY	THE TIME BETWEEN SUNRISE TO SUNSET.
NIGHT	THE TIME BETWEEN SUNSET TO SUNRISE
CHANGE	TO GIVE A DIFFERENT FORM OR APPEARANCE TO
TEMPERATURE	HOW HOT OR COLD SOMETHING IS
THERMOMETER	A TOOL USED TO MEASURE TEMPERATURE

MOTION	
MOTION	A CHANGE IN POSITION
DIRECTION	THE PATH WAY OF AN OBJECT IN MOTION
ABOVE	ON TOP OF; HIGHER THAN
BELOW	UNDER; UNDERNEATH
BEHIND	AT THE BACK OF; LAST
BESIDE	NEXT TO
STRAIGHT	NOT BENT OR CURVY
ZIGZAG	MOVING WITH SHARP TURNS
BACK AND FORTH	MOVING SIDE TO SIDE
ROUND AND ROUND	MOVING IN A CIRCLE
PROPERTIES OF MATTER	
WEATHER	THE AIR OUTSIDE OBSERVED OVER A SHORT PERIOD OF TIME
SEASONS	WINTER, SPRING, SUMMER, FALL; THERE ARE 4 SEASONS IN 1 YEAR
HEAT	A FORM OF ENERGY
THERMOMETER	A TOOL USED TO MEASURE TEMPERATURE
TEMPERATURE	THE AMOUNT OF HEAT IN MATTER
CALENDAR	A TOOL USED FOR SHOWING THE MONTHS, WEEKS, AND DAYS IN AT LEAST ONE SPECIFIC YEAR
WIND SOCK	A TOOL USED TO SHOW THE DIRECTION OF THE WIND
WEATHER PATTERNS	WHEN WEATHER REPEATS ITSELF
LIQUID	A STATE OF MATTER THAT HAS A DEFINITE VOLUME BUT TAKES THE SHAPE OF
CAREERS	
ASTRONOMER	SCIENTIST THAT STUDIES CELESTIAL PHENOMENA
CONTRACTOR	A GENERAL CONTRACTOR IS RESPONSIBLE FOR THE DAY-TO-DAY OPERATIONS OF A CONSTRUCTION SITE. CONTRACTORS MANAGE SALES PEOPLE, BUILDERS, PLUMBERS, ELECTRICIANS, AND ALL THE OTHER TRADES REQUIRED TO COMPLETE A PROJECT. THE CONTRACTOR ALSO COMMUNICATES INFORMATION TO ALL THE PEOPLE INVOLVED IN THE PROJECT.
DESIGNER	INDUSTRIAL DESIGNERS DEVELOP CONCEPTS AND SPECIFICATIONS THROUGH COLLECTION, ANALYSIS AND SYNTHESIS OF DATA GUIDED BY THE SPECIAL REQUIREMENTS OF THE CLIENT OR MANUFACTURER. DESIGNERS MAKE CLEAR AND CONCISE RECOMMENDATIONS THROUGH DRAWINGS, VERBAL DESCRIPTIONS, AND MATH MODELS.
ENGINEER	PROFESSIONAL THAT APPLIES PRINCIPLES OF SCIENCE AND MATHEMATICS BY WHICH THE PROPERTIES OF MATTER AND THE SOURCES OF ENERGY IN NATURE ARE MADE USEFUL TO PEOPLE.

COMMON CORE
DO OR DON'T. THERE IS NO "I CAN."

A famous resident of Dagobah once said, "Do or do not. There is no try." Jedi Master Yoda may not sit with students in today's classrooms, but modern educational pedagogy would benefit from the concept. The use of "I CAN" statements is bothersome in that knowledge unshared does not contribute to community or society.

The "I can" is unnecessary. Let's help students focus on clear statements of ability. Action versus ability leaves little room for question. Goals are clear-cut, and students take pride in achievement. Visit the Ten80 Elementary wiki site to download pages that help students take charge of their own learning goals. *(available December 2014)*

I count.
I read.
I name.
I write.
I participate.
I understand.

K CCSS Mathematics		Counting & Cardinality	
Indicator		Date Taught	Date As-
Know number names and the count sequence.			
CCSS.MATH.CONTENT.K.CC.A.1	I count to 100 by ones and tens.		
CCSS.MATH.CONTENT.K.CC.A.2	I count forward starting at any number I have learned.		
CCSS.MATH.CONTENT.K.CC.A.3	I write numbers from 0 to 20.		
CCSS.MATH.CONTENT.K.CC.A.3	I write a number to tell about a group of 0 to 20 things.		
Count to tell the number of objects.			
CCSS.MATH.CONTENT.K.CC.A.4	I understand how number names go with counting things in the right order.		
CCSS.MATH.CONTENT.K.CC.A.4.A	I name the number for each thing in a group as I count them.		
CCSS.MATH.CONTENT.K.CC.A.4.B	I understand that the last thing I count tells the number of things in a group.		
CCSS.MATH.CONTENT.K.CC.A.4.B	I know the number of items in a group won't change even when items move.		
CCSS.MATH.CONTENT.K.CC.A.S	I count up to 10 to tell how many things are in a group.		

K CCSS Mathematics Base Ten

Indicator		Date Taught	Date Assessed
Work with numbers 11-19 to gain foundations for place value.			
CCSS.MATH.CONTENT.I.NBT.A.1	I count up to 20 starting at any number under 20.		
CCSS.MATH.CONTENT.I.NBT.A.1	I read and write my numbers to show how many objects are in a group (up to 20).		
Understand place value.			
CCSS.MATH.CONTENT.I.NBT.B.2	I tell how many tens and how many ones are in a number.		
CCSS.MATH.CONTENT.I.NBT.B.2.A	I show that I know what a "ten" is.		
CCSS.MATH.CONTENT.I.NBT.B.2.B	I show that any number between 11 and 19 is a group of ""ten"" and a certain number of ones.		
CCSS.MATH.CONTENT.I.NBT.B.2.C	I show that I understand the numbers I use when I count by tens, have a certain number of tens and 0 ones.		
CCSS.MATH.CONTENT.I.NBT.B.3	I compare two-digit numbers using <, =, and > because I understand tens and ones.		
Use place value understanding and properties of operations to add and subtract.			
CCSS.MATH.CONTENT.I.NBT.C.4	I use math strategies to help me solve and explain addition problems within 100.		
CCSS.MATH.CONTENT.I.NBT.C.4	I use objects and pictures to help me solve and explain addition problems within 100.		
CCSS.MATH.CONTENT.I.NBT.C.4	I add ones and tens when adding 2-digit numbers.		
CCSS.MATH.CONTENT.I.NBT.C.4	Sometimes I have to compose a group of higher number when adding 2-digit numbers.		
CCSS.MATH.CONTENT.I.NBT.C.S	I find 10 more or 10 less in my head.		
CCSS.MATH.CONTENT.I.NBT.C.6	I use different strategies to subtract multiples of 10 (10-90) from numbers under 100. I write the matching number sentence and explain my strategy.		

CCSS Mathematics — Operations & Algebraic Thinking K

Indicator		Date Taught	Date Assessed
Understand addition, and understand subtraction.			
CCSS.MATH.CONTENT.K.OA.A.1	I use what makes sense to me to show that I know how to add.		
CCSS.MATH.CONTENT.K.OA.A.1	I use what makes sense to me to show that I know how to subtract.		
CCSS.MATH.CONTENT.K.OA.A.2	I use objects or drawings to show that I can solve addition word problems up to 10.		
CCSS.MATH.CONTENT.K.OA.A.2	I use objects or drawings to show that I can solve subtraction word problems up to 10.		
CCSS.MATH.CONTENT.K.OA.A.3	I take apart any number from 1 to 10 to show that I understand that number. (5 = 2 + 3)		
CCSS.MATH.CONTENT.K.OA.A.4	I take any number from 1 to 9 and show what I need to add to it to make 10.		
CCSS.MATH.CONTENT.K.OA.A.5	I add numbers within 5.		
CCSS.MATH.CONTENT.K.OA.A.5	I subtract numbers within 5.		

CCSS Mathematics — Base Ten K

Indicator		Date Taught	Date Assessed
Work with numbers 11-19 to gain foundations for place value.			
CCSS.MATH.CONTENT.K.NBT.A.1	1 make and take apart numbers from 11 to 19 by telling how many tens and ones are in the number.		
CCSS.MATH.CONTENT.K.NBT.A.1 1 can show how many tens and ones in numbers from 11 to 19 by drawing a picture or writing a number sentence.	1 show how many tens and ones in numbers from 11 to 19 by drawing a picture or writing a number sentence.		

CCSS Mathematics K — Measurement & Data

Indicator		Date Taught	Date Assessed
Describe and compare measurable attributes.			
CCSS.MATH.CONTENT.K.MD.A.1			
CCSS.MATH.CONTENT.K.MD.A.2			
Classify objects and count the number of objects in each category.			
CCSS.MATH.CONTENT.K.MD.A.3			
CCSS.MATH.CONTENT.K.MD.A.3			

CCSS Mathematics K — Geometry

Indicator		Date Taught	Date Assessed
Identify and describe shapes.			
CCSS.MATH.CONTENT.K.G.A.1 I can name and tell about shapes I see around me.	I name and tell about shapes I see around me.		
CCSS.MATH.CONTENT.K.G.A.1 I can tell where I see shapes by using words like: above, below, beside, in front of, behind and next to.	I tell where I see shapes by using words like: above, below, beside, in front of, behind and next to…		
CCSS.MATH.CONTENT.K.G.A.2	I name shapes no matter how big they are or which way they are turned.		
CCSS.MATH.CONTENT.K.G.A.3	I tell if a shape is two-dimensional (flat) or three-dimensional (solid).		
Analyze, compare, create and compose shapes.			
CCSS.MATH.CONTENT.K.G.A.4	I think about and compare two-dimensional and three-dimensional shapes.		
CCSS.MATH.CONTENT.K.G.A.5	I make shapes by drawing them or by using things like sticks and clay.		
CCSS.MATH.CONTENT.K.G.A.6	I use simple shapes to make larger shapes.		

CCSS Reading Foundations

Indicator		Date Taught	Date Assessed
Print Concepts			
CCSS.ELA-Literacy.RF.K.l	I show that I know how books should be read.		
CCSS.ELA-Literacy.RF.K.l.A	I read the words in a book in the right order.		
CCSS.ELA-Literacy.RF.K.l.B	I understand that words I say can be written using letters in a certain order.		
CCSS.ELA-Literacy.RF.K.l.C	I understand that words have spaces between them.		
CCSS.ELA-Literacy.RF.K.l.D	I name all of my upper and lower case letters in the alphabet.		
CCSS.ELA-Literacy.RF.K.2	I show that I know how words and their parts go together.		
CCSS.ELA-Literacy.RF.K.2.A	I recognize and make rhyming words.		
CCSS.ELA-Literacy.RF.K.2.B	I count and divide words into syllables.		
CCSS.ELA-Literacy.RF.K.2.C	I blend and take apart the beginning sounds and ending parts of one-syllable words.		
CCSS.ELA-Literacy.RF.K.2.0	I find and say the beginning, middle and last sound in simple words.		
CCSS.ELA-Literacy.RF.K.2.E	I make new words by changing a consonant or a vowel sound in a word I already know.		
Phonics and Word Recognition			
CCSS.ELA-Literacy.RF.K.3	I show what I have learned about letters and sounds by reading words.		
CCSS.ELA-Literacy.RF.K.3.A	I say the most common sound for each consonant in the alphabet.		
CCSS.ELA-Literacy.RF.K.3.B	I match long and short vowel sounds with the letters that go with them.		
CCSS.ELA-Literacy.RF.K.3.C	I read common sight words.		
CCSS.ELA-Literacy.RF.K.3.0	I tell the difference between and read similar words by looking at the letters that are different.		
Fluency			
CCSS.ELA-Literacy.RF.K.4	I read and understand books at my level well.		

CCSS Reading — Literature — K

Indicator		Date Taught	Date Assessed
Key Ideas and Details			
CCSS.ELA-LITERACY.RL.K.1	I ask and answer questions about important details in stories.		
CCSS.ELA-LITERACY.RL.K.2	I retell a story I know using important details from the story.		
CCSS.ELA-LITERACY.RL.K.3	I know the characters, setting and what happens in stories.		
Character and Structure			
CCSS.ELA-LITERACY.RL.K.4	I ask and answer questions about new words in stories.		
CCSS.ELA-LITERACY.RL.K.5	I know difference between the different kinds of fiction I read.		
CCSS.ELA-LITERACY.RL.K.6	I name a book's author and illustrator and know what their jobs are.		
Integration of Knowledge and Ideas			
CCSS.ELA-LITERACY.RL.K.7	I know how the words and pictures go together in stories.		
CCSS.ELA-LITERACY.RL.K.9	I know what is the same and different about the actions of characters in stories I know.		
Range of Reading and Level of Text Complexity			
CCSS.ELA-LITERACY.RL.K.10	I am an important part of fiction reading activities in my classroom.		

CCSS Reading — Writing — K

Indicator		Date Taught	Date Assessed
Text Types and Purposes			
CCSS.ELA-LITERACY.W.K.1	I draw or write to help share what I think.		
CCSS.ELA-LITERACY.W.K.2	I draw or write to explain a topic.		
CCSS.ELA-LITERACY.W.K.3	I draw or write to tell an organized story about something that has happened.		
Production and Distribution of Writing			
CCSS.ELA-LITERACY.W.K.5	I listen to my friends' ideas to help add details to my stories.		
CCSS.ELA-LITERACY.W.K.6	I use technology to share my ideas.		
Research to Build and Present Knowledge			
CCSS.ELA-LITERACY.W.K.7	I help my class learn about a subject, and then I write about it.		
CCSS.ELA-LITERACY.W.K.8	I use what I know and have read to answer questions.		

CCSS Reading K — Informational Text

Indicator		Date Taught	Date Assessed
Key Ideas and Details			
CCSS.ELA-LITERACY.RI.K.I	I ask and answer questions about important details in nonfiction books.		
CCSS.ELA-LITERACY.RI.K.2	I name the main topic and important details i n nonfiction books.		
CCSS.ELA-LITERACY.RI.K.3	I tell how people, events or ideas are connected.		
Craft and Structure			
CCSS.ELA-LITERACY.RI.K.4	I ask and answer questions about new words in nonfiction books.		
CCSS.ELA-LITERACY.RI.K.S	I find the front cover, back cover and title page in nonfiction books.		
CCSS.ELA-LITERACY.RI.K.6 I	I name the main topic and important details i n nonfiction books.		
Integration of Knowledge a nd Ideas			
CCSS.ELA-LITERACY.RI.K.7	I tell how the words and pictures go together in nonfiction books.		
CCSS.ELA-LITERACY.RI.K.8	I find the reasons an author gives to make the information more clear.		
CCSS.ELA-LITERACY.RI.K.9	I can tell how two nonfiction books about the same thing a re al ike and different.		
Range of Reading and Level of Text Complexity			
CCSS.ELA-LITERACY.RI.K.IO	I am an important part of nonfiction reading activities in my classroom.		

CCSS Reading K — Speaking & Listening

Indicator		Date Taught	Date Assessed
Comprehension and Collaboration			
CCSS.ELA-LITERACY.SL.K.I	I show that I know how to have good conversations with my friends and teachers.		
CCSS.ELA-LITERACY.SL.K.I.A	I listen and take turns in conversations.		
CCSS.ELA-LITERACY.SL.K.I.B	I have long conversations with another person.		
CCSS.ELA-LITERACY.SL.K.2	I can talk about the important ideas in a story.		
CCSS.ELA-LITERACY.SL.K.3	I ask and answer questions about what I have heard.		
Presentation of Knowledge and Ideas			
CCSS.ELA-LITERACY.SL.K.4	I use details when I tell about people, places and things.		
CCSS.ELA-LITERACY.SL.K.S	I draw to add details to what I share.		
CCSS.ELA-LITERACY.SL.K.6	I speak and share my ideas clearly.		

CCSS Reading — Language — K

Indicator		Date Taught	Date Assessed
Conventions of Standard English			
CCSS.ELA-LITERACY.L.K.l	I show that I know how to use words correctly when I write and speak.		
CCSS.ELA-LITERACY.L.K..Al	I print lots of upper and lowercase letters.		
CCSS.ELA-LITERACY.L.K..Bl	I use nouns (words that name) and verbs (action words).		
CCSS.ELA-LITERACY.L.K..Cl	I make nouns plural (more than one) by adding "s" or "es" to the end.		
CCSS.ELA-LITERACY.L.K..Ol	I understand and use question words. (who, what where,when,why,how)		
CCSS.ELA-LITERACY.L.K..El	I use common prepositions (to,from,in, out,on, off,for,of,by,with).		
CCSS.ELA-LITERACY.L.K..Fl	I create longer complete sentences with my class.		
CCSS.ELA-LITERACY.L.K.2	I show that I know how to write sentences correctly.		
CCSS.ELA-LITERACY.L.K.2.A	I capitalize the first word in a sentence.		
CCSS.ELA-LITERACY.L.K.2.A	I capitalize the word "1".		
CCSS.ELA-LITERACY.L.K.2.B	I find and name punctuation at the end of a sentence.		
CCSS.ELA-LITERACY.L.K.2.C \	I write a letter or letters for most consonant sounds.		
CCSS.ELA-LITERACY.L.K.2.C	I write a letter or letters for most short vowel sounds.		
CCSS.ELA-LITERACY.L.K.2.0	I use what I know about letters and sounds to spell easy words.		
Vocabualry Acquisition and Use			
CCSS.ELA-LITERACY.L.K.4	I figure out what words mean by thinking about what I have read.		
CCSS.ELA-LITERACY.L.K.4.A	I find new meanings for words I already know and use them correctly.		
CCSS.ELA-LITERACY.L.K.4.B	I use beginnings and endings of words to help me figure out what they mean (-ed,-s,re-,un-,pre-,-ful,-less).		
CCSS.ELA-LITERACY.L.K.S	I figure out how words are related. I figure out how meanings might be alike.		
CCSS.ELA-LITERACY.L.K.S.A	I sort things into groups. I use groups to understand things better.		
CCSS.ELA-LITERACY.L.K.S.B	I match some action verbs and adjectives with their opposites.		
CCSS.ELA-LITERACY.L.K.S.C	I tell how words are used in real-life.		
CCSS.ELA-LITERACY.L.K.S.O	I tell the difference between verbs (action words) that are almost alike.		
CCSS.ELA-LITERACY.L.K.6	I use new words in different ways to show that I know what they mean.		

RHOMBI'S ADVENTURES IN 3D
WORKS CITED

"Circle." Wikipedia, the Free Encyclopedia. Wikimedia Foundation, Inc, n.d. Web. 7 July 2014.

Dehaene, Stanislas. The Number Sense: How the Mind Creates Mathematics. New York: Oxford UP, 1997. Print.

Devlin, Keith J. The Math Gene: How Mathematical Thinking Evolved and Why Numbers Are Like Gossip.

    New York: Basic Books, 2000. Print.

Ferris Bueller's Day Off. Dir. John Hughes. Perf. Matthew Broderick, Mia Sara, Alan Ruck, Jennifer Grey,

    and Jeffrey Jones. Paramount, 1986. Film.

Feynman, Richard P, and Jeffrey Robbins. The Pleasure of Finding Things Out:

    The Best Short Works of Richard P. Feynman. Cambridge: Perseus Books, 1999. Print.

Furoy, Michael, Rosemary T. Wong, and Harry K. Wong. The Effective Teacher. Mountain View:

    Harry K. Wong Publications, Inc., 2009. Print.

Ifrah, Georges. From One to Zero: A Universal History of Numbers. New York: Viking, 1985. Print.

Lee, Joon S., and Herbert P. Ginsburg. "Preschool Teachers' Beliefs About Appropriate Early Literacy and

    Mathematics Education for Low and Middle-Socioeconomic Status Children." Early Education

    and Development 18.1 (2007):   111-143. Print.

Ma, Liping. Knowing and Teaching Elementary Mathematics: Teachers' Understanding of Fundamental

    Mathematics in China and the United States. Hillsdale, NJ: Lawrence Erlbaum Associates, 1999. Print.

Mathemagic. Chicago: World Book—Childcraft International, Inc., 1980. Print.

McLeish, John. The Story of Numbers. New York: Fawcett Columbine, 1994. Print,

Next Generation Science Standards. N.p., n.d. Web. 7 July 2014.

Wassily Kandinsky – biography, paintings, books. N.p., n.d. Web. 7 July 2014.

"Inquiry Based Science: What Does It Look Like??" Connect Magazine. 1995: 13. Print.